Advances in Intelligent Systems and Computing

Volume 831

Series editor

Janusz Kacprzyk, Polish Academy of Sciences, Warsaw, Poland
e-mail: kacprzyk@ibspan.waw.pl

The series "Advances in Intelligent Systems and Computing" contains publications on theory, applications, and design methods of Intelligent Systems and Intelligent Computing. Virtually all disciplines such as engineering, natural sciences, computer and information science, ICT, economics, business, e-commerce, environment, healthcare, life science are covered. The list of topics spans all the areas of modern intelligent systems and computing such as: computational intelligence, soft computing including neural networks, fuzzy systems, evolutionary computing and the fusion of these paradigms, social intelligence, ambient intelligence, computational neuroscience, artificial life, virtual worlds and society, cognitive science and systems, Perception and Vision, DNA and immune based systems, self-organizing and adaptive systems, e-Learning and teaching, human-centered and human-centric computing, recommender systems, intelligent control, robotics and mechatronics including human-machine teaming, knowledge-based paradigms, learning paradigms, machine ethics, intelligent data analysis, knowledge management, intelligent agents, intelligent decision making and support, intelligent network security, trust management, interactive entertainment, Web intelligence and multimedia.

The publications within "Advances in Intelligent Systems and Computing" are primarily proceedings of important conferences, symposia and congresses. They cover significant recent developments in the field, both of a foundational and applicable character. An important characteristic feature of the series is the short publication time and world-wide distribution. This permits a rapid and broad dissemination of research results.

More information about this series at http://www.springer.com/series/11156

Katarzyna Arkusz · Romuald Będziński
Tomasz Klekiel · Szczepan Piszczatowski
Editors

Biomechanics in Medicine and Biology

Proceedings of the International Conference of the Polish Society of Biomechanics, Zielona Góra, Poland, September 5–7, 2018

 Springer

Editors
Katarzyna Arkusz
Division of Biomedical Engineering,
  Department of Mechanical Engineering
University of Zielona Góra
Zielona Góra, Poland

Romuald Będziński
Division of Biomedical Engineering,
  Department of Mechanical Engineering
University of Zielona Góra
Zielona Góra, Poland

Tomasz Klekiel
Division of Biomedical Engineering,
  Department of Mechanical Engineering
University of Zielona Góra
Zielona Góra, Poland

Szczepan Piszczatowski
Division of Biocybernetics and Biomedical
  Engineering, Department of Mechanical
  Engineering
Białystok University of Technology
Białystok, Poland

ISSN 2194-5357   ISSN 2194-5365   (electronic)
Advances in Intelligent Systems and Computing
ISBN 978-3-319-97285-5   ISBN 978-3-319-97286-2   (eBook)
https://doi.org/10.1007/978-3-319-97286-2

Library of Congress Control Number: 2018949384

This Springer imprint is published by the registered company Springer Nature Switzerland AG
The registered company address is: Gewerbestrasse 11, 6330 Cham, Switzerland

Preface

The field of biomechanics and biomedical engineering is constantly changing, and every year, scientists try to apply physics and other mathematically based forms of analysis to discover the limits and capabilities of biological systems. Biomechanics and new technologies are crucial to providing human health and delivering new solutions in medicine, especially with respect to the problems related to an ageing population. In high-performance sports, practitioners employ a variety of tests and tools, mainly to predict performance and flag when an injury will occur.

We proudly present the Proceedings of the **International Conference of the Polish Society of Biomechanics "BIOMECHANICS 2018"**, which was held in Zielona Góra from 5 to 7 September 2018, which discusses recent research innovations in biomechanics. The conference was organized by the Polish Society of Biomechanics and Biomedical Engineering Division of the University of Zielona Góra. This conference has been held in Poland on a regular basis since 1981. The primary goal of the conference is to provide an opportunity for participants from all corners of the world to share and exchange ideas and experiences relating to biomechanics. The "BIOMECHANICS 2018" conference was a great opportunity for an exchange of ideas and presentations of the latest developments in all the areas within the field of biomechanics including biofluid mechanics, biomaterials and biosensors, biotribology, cardiovascular biomechanics, cellular and molecular biomechanics, clinical biomechanics, computational imaging and simulation technologies in biomechanics, human movement, impact/injury biomechanics, musculoskeletal biomechanics, neuro-biomechanics, rehabilitation biomechanics, soft and hard tissue biomechanics, spine, sport biomechanics and technology.

This volume includes 26 papers accepted for publication from among those submitted on the "BIOMECHANICS 2018" conference. Aiming for the high scientific merit of the meeting and international recognition of the proceedings, all submissions were subjected to a thorough peer review process (minimum of two independent reviews per paper) and only those with consistent and strong recommendations from the reviewers were accepted. In this proceedings volume, the accepted papers are organized into five sections:

- Clinical Biomechanics
- Computational Imaging and Simulation Technologies in Biomechanics
- Impact/Injury Biomechanics
- Soft and Hard Tissue Biomechanics
- Sport Biomechanics and Technology

The "BIOMECHANICS 2018" conference promotes innovative activities in the field of modern clinical achievements, imaging processing and simulation method activities, allowing for a better understanding of the mechanisms of functioning in human tissues and organs in the interaction with elements of biomedical engineering, i.e. stents, implants and artificial organs. Equally interesting is the behaviour of human tissues under loads caused by excessive effort or rapid overloading in the area of sports. The participants of the "BIOMECHANICS 2018" conference presented many interesting and diverse research and analysis methods to better explain how the human body works, which translates into the development of treatment methods.

We believe that this book will become a great reference tool for scientists working in the areas of biomechanics. Readers are kindly encouraged to contact the corresponding authors for further details about their research.

Katarzyna Arkusz
Romuald Będziński
Tomasz Klekiel
Szczepan Piszczatowski

Organization

The International Conference of the Polish Society of Biomechanics "BIOMECHANICS 2018" was organized by the Polish Society of Biomechanics and Biomedical Engineering Division of the University of Zielona Góra.

BIOMECHANICS 2018 Scientific Committee

General Chair

Romuald Będziński   University of Zielona Góra, Poland

Vice-chair

Szczepan Piszczatowski   Białystok University of Technology, Poland

Ricardo Alves de Sousa, Portugal
Jorge Ambrosio, Portugal
Jan Awrejcewicz, Poland
Wojciech Blajer, Poland
Lajos Borbás, Hungary
Tadeusz Burczyński, Poland
Krzysztof Buśko, Poland
Luca Cristofolini, Italy
Adam Czaplicki, Poland
Jan R. Dąbrowski, Poland
Manuel Doblare, Spain
Lechosław Dworak, Poland
Igor Emri, Slovenia
Stephan Ferguson, Switzerland

Jarosław Filipiak, Poland
Marc Geers, Holland
Sanjay Gupta, India
Marek Gzik, Poland
Paweł Jarmużek, Poland
Grzegorz Juras, Poland
Aleksander Kabsch, Poland
Krzysztof Kędzior, Poland
Natalia Kiziłowa, Poland
Michał Kleiber, Poland
Tomasz Klekiel, Poland
Krzysztof Kwiatkowski, Poland
Damien Lacroix, UK
Tomasz Łodygowski, Poland

Andrzej Maciejczak, Poland
Ewa Majchrzak, Poland
Jerzy Małachowski, Poland
Ryszard Maroński, Poland
Andrzej Mastalerz, Poland
Daniel Matej, Czech Republic
Robert Michnik, Poland
Grzegorz Milewski, Poland
Karol Miller, Australia
Zbigniew Nawrat, Poland
Tadeusz Niezgoda, Poland
Yuriy Nyashin, Russia
Urszula Pasławska, Poland
Celina Pezowicz, Poland

Halina Podbielska, Poland
John Rasmussen, Denmark
Gwendolen Reilly, UK
Alicja Rutkowska-Kucharska, Poland
Patrik Schmuki, Germany
Konstanty Skalski, Poland
Małgorzata Syczewska, Poland
Marek Synder, Poland
Ioan Szava, Romania
Vladimir Tregubov, Russia
Czesław Urbanik, Poland
Andrzej Wit, Poland
Wojciech Wolański, Poland
Daniel Zarzycki, Poland

BIOMECHANICS 2018 Organizing Committee

General Chair

Katarzyna Arkusz   University of Zielona Góra, Poland

Vice-chair

Tomasz Klekiel   University of Zielona Góra, Poland

Secretary

Agnieszka Mackiewicz   University of Zielona Góra, Poland

Agnieszka Kaczmarek-Pawelska
Jagoda Kurowiak
Marta Nycz
Ewa Paradowska
Monika Ratajczak
Rafał Rudyk
Urszula Skiruk

Acknowledgement

We would like to thank all participants, plenary speakers and paper reviewers for their scientific and personal contributions to the BIOMECHANICS 2018 conference.

Many thanks and much appreciation are due to the peer reviewers from Belgium, Czech Republic, Hungary, Italy, Poland, Portugal, Slovakia and the UK who have greatly contributed to a critical selection of the best papers and whose remarks and suggestions have significantly helped the authors improve the quality of their papers.

Ministry of Science and Higher Education

Republic of Poland

International Conference of the Polish Society of Biomechanics - BIOMECHANICS 2018 was financed under contract 777P-DUN/2018 from the resources of Ministry of Science and Higher Education allocatied for activities which disseminate the science.

Contents

Clinical Biomechanics

Human Red Blood Cell Properties and Sedimentation Rate: A Biomechanical Study

Natalya Kizilova[1]([⊠]), Liliya Batyuk[2], and Vitalina Baranets[3]

[1] Warsaw University of Technology,
Nowowiejska street, 24, 00-665 Warsaw, Poland
n.kizilova@gmail.com
[2] Kharkov National Medical University,
Nauki avenue, 2, Kharkiv 61022, Ukraine
[3] Kharkov National University, Svobody square, 4, Kharkiv 61022, Ukraine

Abstract. Human blood is widely used for clinical diagnostics due to its easy accessibility and high sensitivity for any metabolic disorders and diseases. In the paper different mechanical and electric properties of red blood cells (RBC) useful for diagnostics are discussed. The experimental data on the erythrocyte sedimentation rate (ESR) test in its standard procedure and continuous computer assisted assessment are presented. The review of mathematical approaches for reasonable ESR estimation is given. The continuous model of blood as three phase suspension is used for numerical estimation of aggregate ability of RBC. The problem on RBC aggregation and sedimentation in a thin vertical tube is considered. A numerical solution of the formulated hyperbolic problem is obtained by the method of characteristics. Numerical computations have been carried out for a wide range of RBC parameters proper to healthy state and patients with cancer, drug and food allergy. It is shown the continuous sedimentation curves give more diagnostic information than the standard ESR value. Based on the experimental and theoretical study, a biomechanical interpretation of the ESR curves is proposed.

Keywords: Biomechanics · Erythrocyte sedimentation rate
Medical diagnostics

1 Introduction

Blood is a unique tissue circulating between all the cells, tissues and organs of the body and carrying the organic and mineral components as well as products of the tissue metabolism, new growth, cellular apoptosis, and other physiological processes [1]. In that way, blood is the most proper tissue for easy and detailed diagnosis of the body metabolism and state [2]. Any sort of ionic, osmotic or other disease-related changes produce noticeable variations of the membrane surface properties of the red blood cells (RBC) that compose ~ 40–49% of the blood volume [1, 2]. Therefore, the electric (electric surface charge, electric conductivity, dielectric permittivity), mechanical (density, membrane shape, thickness, deformability, fluidity) and other physical

K. Arkusz et al. (Eds.): BIOMECHANICS 2018, AISC 831, pp. 3–22, 2019.
https://doi.org/10.1007/978-3-319-97286-2_1

(adhesiveness) properties of RBC can be used for early diagnostics of pathological processes [2]. One of the most common and easy clinical tests is the erythrocyte sedimentation rate (ESR) test. The ESR value is the height of the clean blood plasma (BP) zone formed over the RBC zone during gravitational sedimentation of RBC in the blood sample placed into a thin vertical tube. There are several distinct test methodology (Wintrobe, Westergren, Panchenkov) with different blood type (arterial or venous), length (10–30 cm) and diameter (1–4 mm) of the tube, and duration of the test (1–3 h) accepted in different countries [2, 3]. The ESR is an excellent index of inflammatory state associated with cancers, infections, and autoimmune diseases like systemic vasculitis, temporal arteritis, polymyalgia rheumatica, and others. Inflammations cause abnormal proteins to appear in blood and change the surface properties of RBC that promotes their aggregation and faster sedimentation [4–6]. Disregarding the long history of the ESR test [7] and its numerous modifications [2], recently published reviews and research papers on novel aspects of anomalous high [8, 9] and low [10] ESR, its applicability to the elderly [9] and young [11] patients, as well as in pregnancy [12] revealed new approaches for the test procedures and data analyses. During the last decades novel approaches for ESR-based diagnosis of hypercholesterolemia [13], heart failure [14], osteomyelitis [15], and other specific diseases have been proposed.

In this paper a short comprehensive review on the RBC physical properties and aggregation ability, and on mathematical models for EBS modelling and computation of the ESR (normal, pathologically fast or short) is given. The results of measurements of ESR in healthy volunteers and patients with drag and food allergy are presented.

1.1 Electric and Mechanical Properties of RBC

Blood as a concentrated suspension of red blood cells (RBC) is characterized by passive electric properties like electric impedance, conductivity, and dielectric permittivity. Electromechanical properties of BP and blood cells (erythrocytes, leucocytes and platelets) determine RBC and platelets interaction and aggregation, blood rheology and blood clot formation, blood flow through the vessels and tubes of biomedical units [1, 2, 16]. Passive electric property is electric resistivity $Z = \delta\phi/I$ that is the ratio of the applied electric potential $\delta\phi$ to the electric current I measured in the sample. Electric resistivity of biological tissues depends on water contents, ionic strength of BP, membrane conductivity and some other biophysical factors. When a constant electric potential $\delta\phi$ is applied to a tissue, the electric current I through it decreases due to generation a counter directed electromotive force E, and the Ohm's law is $Z = (\delta\phi - E)/I$. To avoid the electromotive force induction, the measurements are usually conducted in oscillating electromagnetic fields (EMF), and the complex impedance $\tilde{Z} = Z + iX$, where X is the reactance is measured. The values Z, X for different tissues in healthy state and at different diseases have been measured and classified in biophysical handbooks [2].

The biological tissues are not ideal dielectrics and their dielectric properties are described by complex dielectric permittivity $\varepsilon^* = \varepsilon + i\varepsilon'$, where ε is the real permittivity, $\varepsilon' = \sigma/(2\pi f)$, $\sigma = Z^{-1}$ is electric conductivity, f is the EMF frequency. The ratio

$\alpha_\varepsilon = \mathrm{atan}(\varepsilon'/\varepsilon)$ determines the tangent of dielectric loss due to transfer of some EMF energy into heat and dielectric relaxation [16, 17].

RBC and other cells in aqueous conditions at physiological pH values have negative electric surface charge $\sigma_e = q/S$, where q is the charge of the cell, S is its surface area, produced by dissociation of the membrane proteins, ion absorption and water hydration layers formed at the cellular surfaces [16]. The surface charge of healthy RBC was found within the limits $\sigma_e = 3 \cdot 10^{-3} \div 1.5 \cdot 10^{-2}$ S/m^2. The surface charge can be quantitatively estimated by electrophoretic mobility (EPM) of the cells in external static electric field, which is the ratio of the cell velocity to the strength of the applied field. Healthy RBC demonstrate EPM = 1.1 – 1.3 μm/s/V/cm [16].

The dielectric permittivity of native blood [17, 18] and RBC suspensions [19, 20] have been thoroughly examined for the medical diagnostic purposes. In the external EMF electric and dielectric properties of biological tissues demonstrate dispersion, i.e. dependence of ε and σ on the EMF frequency f: α – dispersion at low frequencies (f = 10 – 10^3 Hz), β – dispersion in the radio wave range (f = 10^4 – 10^8 Hz), and γ – dispersion in the microwave range (f > 10^{10} Hz). In the native blood and RBC suspensions α – dispersion is almost absent; β – dispersion is determined by noticeable decrease in capacity of the cell membranes, and dipole polarization of hemoglobin, membrane and BP proteins; and γ – dispersion is determined by polarization of water molecules, and structure and behavior of the hydration shells of RBC [13, 14]. The ε and σ values for native human blood at low and high frequency limits are $\varepsilon = 7350$, $\sigma = 0.49$ S/m and $\varepsilon = 160$, $\sigma = 0.9$ S/m accordingly [17].

Mechanical properties of RBC (size, shape, deformability, viscoelasticity) are related to their electrical characteristics. Inside the organism blood moves through the blood vessels at a strong influence of electric fields generated by contracting heart, skeletal and smooth muscles, and therefore the electric properties of RBC and BP influence their movement, blood viscosity, and efficiency of the transport of oxygen, carbon dioxide, mineral, organic and other components. Blood viscosity as a main determinant in hemodynamics, depends on RBC concentration, size, shape and deformability, protein concentration in BP, temperature, and shear rate in the flow [1, 16]. In the inflammation state, cancer, necrosis and other severe diseases the fragments of membranes, specific proteins and viruses can modify the RBC membranes, change their hydration shells [20], decrease their surface charge [16] and, therefore, increase their aggregate ability that leads to high ESR [3–6, 8, 9]. When the RBC membranes are overloaded by absorbed particles and cations, cholesterol or other substances their charge is decreased and their membrane become thicker that essentially increases RBC rigidity and decreases their deformability and ability to pass through the capillaries [1, 2].

The most relevant quantitative estimations of the RBC electromechanical properties and aggregate ability for medical diagnostics purposes can be carried out by EPM based estimations of the electric charge [2, 16], microwave dielectric spectroscopy [17–20], and ESR tests [1, 2].

1.2 RBC Aggregation in Health and Pathology

Healthy RBCs demonstrate reversible rouleaux formation, i.e. elongated aggregates composed by chains of RBC stacked together as coins. The aggregates are easily destroyed by shear stress in the blood flow through the vessels or tubes of the external blood circulation/oxygenation systems. In the slow flows and in the stagnant regions of secondary flows the RBC can form more complex aggregates of >100 cells, thus, hampering the circulation and increasing the RBC sedimentation, because larger aggregates settle faster in the gravity field.

Microscopic methods for estimation of RBC aggregation rate are based on direct counting of the average number of cells in the chains on the dry smears (smeared drop of blood deposited on a glass slide) or on micro images of diluted blood [21, 22]. Optical methods based on light scattering patterns in the colloidal systems and suspensions are also used for the RBC aggregation rate estimation [21, 23].

According to electron microscope images, there are narrow gaps between the flattened sides of RBC, and the width of the gap corresponds to the length of a bridge molecule, which is fibrin in normal blood, and other polymer molecule at experimental conditions (polylysine, dextrans, etc.) [21]. The bigger number of bridges provides stronger binding of RBC in the aggregates. Adhesions of the bridge molecules at RBC surfaces are reversible and can be destroyed by shear stresses and chemical influences [22]. At physiological pH when RBC have negative surface charge and the bridge molecules have positive charged ends, aggregation is determined by the electrostatic forces, while for neutral or negatively charges ends it is determined by van-der-Vaals forces and hydrogen binds which are less stable for deformations. In that way, the microphysical theories of RBC aggregation are based on physics of double electric layers and hydrate shells, electric and viscoelastic interactions of cellular surfaces and bridge molecules [23, 24].

The Monte-Carlo simulations of geometry and fractal properties of the aggregates composed by absorption of particles with determined electric potentials revealed that aggregation is a very complex phenomena even in binary mixtures and needs additional experimental and theoretical studies [25]. RBC aggregation is much more complex process that is influenced by pH of blood, cell shape, electric and adhesive properties, and deformability. Different medicines and toxins produced by microorganisms and viruses cause various changes in the mechanical and electric properties of RBC and BP. The simplest microphysical model of RBC aggregation is based on Newton dynamic law [26]

$$m_i \frac{d\bar{v}_i}{dt} = \sum_{i \neq j} \bar{f}_{ij} + \bar{f}_i^h, \qquad (1)$$

where m_i is the mass of i-th particle, $\bar{v}_i$ is its velocity, $\bar{f}_{ij}$ is the force acting on i-th particle from j-th particle, $\bar{f}_i^h$ is the hydrodynamic force.

Generally speaking, the forces $\bar{f}_{ij}$ could be divided into attraction forces $\bar{f}_{ij}^a$ promoting particle approaching, interaction and binding, and repulsive forces $\bar{f}_{ij}^r$. One of the repulsive forces is the elastic repulsion $\bar{f}_{ij}^e$ determined by physical occupation of space by the contacting particles, their elastic deformation and interaction as viscoelastic fluid-filled charged shells. Then $\bar{f}_{ij} = \bar{f}_{ij}^a + \bar{f}_{ij}^e$ where

$$\vec{f}_{ij}^{\,e} = \begin{cases} k\left(r_i + r_j - d_{ij}\right)^{3/2}\bar{n}_{ij} & \text{if } \begin{array}{l} d_{ij} \le r_i + r_j \\ d_{ij} > r_i + r_j \end{array} \end{cases}, \tag{2}$$

where $r_{i,j}$ are radiuses of the i-th and j-th particles interacting, d_{ij} is the distance between their physical centers, $\bar{n}_{ij}$ is the unit vector directed from j-th to i-th particle. Interaction of two particles depends on their Morse potentials

$$\varPhi_{ij} = D_{ij}\left(e^{2B\left(\delta_0 - \delta_{ij}\right)} - 2e^{B\left(\delta_0 - \delta_{ij}\right)}\right),$$

where D_{ij} are the coefficients describing surface properties, $\delta_{ij} = \left(d_{ij} - r_i - r_j\right)$ are the distances between the surfaces of the interacting particles, $\delta = \delta_0$ if $d_{ij} < r_i + r_j$, B is the scalar coefficient which is inverse to the double electric layer thickness.

Then the attraction force can be written as

$$\vec{f}_{ij}^{\,a} = -\frac{\partial \varPhi_{ij}}{\partial \delta_{ij}}A = 2D_{ij}AB\left(e^{2B\left(\delta_0 - \delta_{ij}\right)} - e^{B\left(\delta_0 - \delta_{ij}\right)}\right)\bar{n}_{ij}, \tag{3}$$

where A is the area of the contact surface of the particles.

The values B, δ_0, D_{ij} can be taken as computed for ideal charged surfaces and corrected then according to experimental data. Hydrodynamic force is usually estimated from the Stokes law for rigid particle and slow flow at low Reynolds numbers

$$\vec{f}_i^{\,h} = k\mu r_i(\vec{v}_i - \vec{v}_\infty), \tag{4}$$

where μ is the fluid viscosity, $\vec{v}_i$ and $\vec{v}_\infty$ are the velocities of the particle and undisturbed flow, k is the shape coefficient that is $k = 6\pi$ for a sphere.

For the concentrated suspensions $(C > 5\%)$ $\mu = \mu(C, \dot{\gamma})$, where C is the concentration of particles, $\dot{\gamma}$ is the flow shear rate. More detailed models account for nonsteady flow, inertia, Magnus effect, and the following expression can be used instead of (10) [24]:

$$\vec{f}_i^{\,h} = k\mu r_i(\vec{v}_i - \vec{v}_\infty) + k_1 \mu \varDelta \vec{v}_\infty - k_2 \rho r_i^3 \frac{d}{dt}(\vec{v}_i - \vec{v}_\infty)$$
$$+ k_3 \rho r_i^3(\vec{\omega}_i - \vec{\omega}_\infty) \times (\vec{v}_i - \vec{v}_\infty) + k_4\sqrt{\mu\rho}r_i^2 \int_{-\infty}^{t} \frac{d}{d\tau}(\vec{v}_i - \vec{v}_\infty)\frac{d\tau}{\sqrt{t - \tau}}, \tag{5}$$

where $\vec{\omega}_i$ and $\vec{\omega}_\infty$ are angular velocities of the i-th particle and fluid, k_{1-4} are particle specific coefficients.

Numerical computations on the model (1)–(5) revealed significant variations in the RBC aggregation rate produced by particle shape coefficient, concentration, electric charge and rigidity [24]. The presented microrheological model of RBC movement, interaction and aggregation will be used in the next chapters for the continual modeling of blood as a multiphase fluid.

1.3   Continuous Modeling of RBC Aggregation and Sedimentation

In the continual mechanics approach the blood is considered as a multiphase suspension of free RBC (1) and RBC aggregates (2) suspended in the BP (3). During the aggregation and disaggregation the solid phases exchange their mass, momentum and energy when free RBC pass on from the phase 1 into phase 3 and vice versa. Mass, momentum and energy balance equations can be obtained from thermodynamics of the multiphase continua. The BP is usually modeled as an incompressible Newtonian liquid [27, 28]. In the single phase approach the mass and momentum balance equations have the simplest form

$$\text{div}(\vec{v}) = 0, \rho \frac{d\vec{v}}{dt} = -\nabla p + \text{div}(2\mu \hat{e}), \tag{6}$$

and must be completed by the equation of aggregation kinetics

$$\frac{dN}{dt} = \Gamma^+(N, I_e) - \Gamma^-(N, I_e), \tag{7}$$

where N is the number of aggregates per unit volume, $\vec{v}$ is the velocity, p is the hydrostatic pressure, $\hat{e}$ is the strain rate tensor, I_e is the second invariant of $\hat{e}$, ρ is density, $\mu = \mu(I_e)$ is the dynamic viscosity, $\Gamma^{\pm}$ are the aggregation and disaggregation rates accounting for different physical mechanisms (4).

Expression for Γ can be reconstructed from the microrheological theory (1)–(5) or by dimension analysis. For instance, a good approximation for the quasi-steady pulsatile flow of thixotropic liquid in the cylindrical tube can be obtained by assuming [27, 28]

$$\Gamma = -\alpha^+ I_e N_0 n^{1-\gamma^+} + \alpha^- I_e N_0 n^{1-\gamma^-}(1-n)\chi(1-n)f\left(\frac{2\mu I_e}{\sigma*}\right), \tag{8}$$

where $n = N/N_0$ is the dimensionless numerical concentration of the aggregates, N_0 is their initial concentration, $\alpha^{\pm} = \text{const} > 0$, $\gamma^{\pm}$ are shape coefficients ($\gamma^{\pm} = 0$ for spherical particles), $\sigma*$ is the critical stress destroying the aggregates, χ is the Heviside function, f is a monotonous non-decreasing dimensionless function.

Different types of flows of thixotropic suspensions of aggregating particles in tubes and channels, flow stability and aggregation stability have been studied based on the model (6)–(8) [28, 29].

The more detailed model accounting for liquid capture inside the aggregates followed by its progressive squeezing out from the aggregates, i.e. the exchange between the phases 2 and 3, has been developed in [27]. The model is based on the mass and momentum balance Eq. (6), while the following balance equations for numerical (N), mass (C) and volumetric (H) concentrations of the aggregates are accepted instead of (8)

$$\frac{dN}{dt} = -\text{div}\,\vec{J}_N + \Gamma_N, \frac{dH}{dt} = -\text{div}\,\vec{J}_H + \Gamma_H, \frac{dC}{dt} = -\text{div}\,\vec{J}_C, \tag{9}$$

where Γ_H is the BP capture inside the aggregates, $\vec{J}_{N,H,C}$ are the corresponding diffusion fluxes; according to reversible thermodynamic approach they are linear functions of the corresponding thermodynamic forces, namely

$$\vec{J}_H = -D_{HH}\nabla H - D_{HC}\nabla C - \frac{D_{HN}}{N}\nabla N - \frac{D_{HI}}{I_e}\nabla I_e. \tag{10}$$

Some simplified problem formulations based on (6), (9), (10) and analysis of the model parameters were given in [28, 30, 31].

2 Materials and Methods

The experimental study has been carried out in the Allergy Lab of Kharkov Institute of Dermatology (Ukraine). The native venous blood samples were collected from 20 patients (10 males and 10 females; average age 59 ± 9 years) and split into small portions $V = 1$ ml. One portion was studied as a control test while into other portions small amounts of the drugs prescribed to the patient by his/her doctor were added. All the portions have been placed in the standard glass tubes ($d = 1$ mm, $L = 100$ mm) and microtubes ($d = 0.5$ mm, $L = 50$ mm). The tubes were placed in a rack in a vertical position for 3 h at the room temperature. The height h of the transparent layer of blood plasma in the upper part of each tube has been measured each 10 min. As a result of the study the sedimentation curves h(t) have been obtained for each tube. The standard ESR value was computed at t = 1 h of sedimentation.

The curves have been smoothed by a Bayesian filter and the time derivative curves $h'(t)$ have been computed. The typical results of the simultaneous ESR test for the blood of the same patient in the standard tubes and the microtubes are presented in Fig. 1. Thick solid lines in Fig. 1a–d correspond to the control sample and the ESR <10 mm indexes correspond to healthy individuals [2]. The addition of the potential allergen influences on the sedimentation rate by some acceleration or deceleration of the ESR process. Since ESR is determined by the aggregate ability of the RBC in the sample, that means a direct influence on the allergen on the RBC surface. The cases with no allergy for the added drugs are presented in Fig. 1a–d by thin solid lines, while the cases with the confirmed allergy denoted by dashed lines.

The curves $h'(t)$ exhibit similar dynamics (Fig. 1b) with the noticeable maximum at $t = t_{max}$, which corresponds to the maximal velocity of sedimentation followed then by some decrease in the ESR. The decrease in the ESR is caused by the influence of the settled aggregates accumulated at the bottom of the sedimentation tube [33]. The time t_{max} was shown to be a good diagnostic index, which almost independent on the initial concentration of RBC and BP viscosity, and reflects the RBC aggregation rate only [24].

When ESR test is carried out in a microcapillary, a smaller amount of blood is needed and the behavior of the ESR curves in the standard (Fig. 1a) and micro (Fig. 1c) tubes is similar. The difference between the control sample and the sample affected by potential allergens were more distinct in the microtubes (Fig. 1c). The maximal velocity was reached in the microtubes well earlier (t = 35 min) and the dispersion was more noticeable. The peaks in the $h'(t)$ curves in the microcapillary are more acute, i.e. the corresponding time intervals are shorter that allows more precise ESR estimation from the curves. The secondary peaks in ESR at t = 80 min have been detected in the microcapillaries only (Fig. 1d). In that way, when t_{max} is used as a

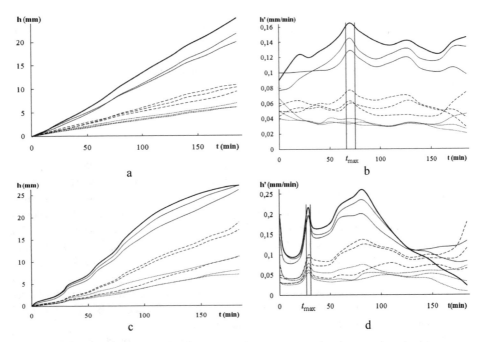

Fig. 1. Experimental curves h(t) (a, c) and h′(t) (b,d) in the standard tubes (a, b) and the microtubes (c, d); solid lines correspond to the control, dashed and dotted lines – to two different potential allergens (medicines).

diagnostic index instead of ESR, the test in the microtubes gives the results earlier, in 30–40 min of the sedimentation process. Thus, the diagnostic information here can be obtained faster, more accurate, and by using a smaller portion of the blood when the microtubes are used.

The results of 20 simultaneous ESR tests with the blood of the same patient conducted in the microtubes are presented in Fig. 2. The variety of the dynamics can be explained by the instability of sedimentation of a suspension of aggregating particles

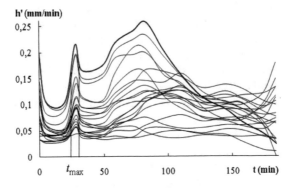

Fig. 2. Measured h′(t) curves and t_{max} location for 20 native blood samples of the same patient.

[24]. Anyway, for all the curves t_{max} values are located in quite narrow limits providing a high exactness in diagnosis of the drug allergy from the ESR tests in the microtubes.

3 Mathematical Problem Formulation

The mixture model of the blood as a three-phase continuous media composed of the free BP (phase 1), RBC aggregates (phase 2) and BP captured inside the aggregates (phase 3) includes the mass and momentum balance equations, and aggregate kinetics equations [27, 28]

$$\frac{\partial \rho^\alpha}{\partial t} + \mathrm{div}(\rho^\alpha \vec{v}^\alpha) = \theta^\alpha,$$
$$\rho^\alpha \frac{d\vec{v}^\alpha}{dt} = -\mathrm{div}(\hat{P}^\alpha) + \rho^\alpha \vec{f}^\alpha + \vec{R}^\alpha, \tag{11}$$
$$\frac{\partial N}{\partial t} + \mathrm{div}(N\vec{v}^2) = \Gamma,$$

where $\alpha = 1, 2, 3$ is the phase number, ρ^α are the densities of the phases, θ^α and $\vec{R}^\alpha$ are the mass and momentum exchange due to the aggregation and BP capture, $\hat{P}^\alpha$ are the stress tensors, $\vec{f}^\alpha$ are the external forces, Γ is the RBC aggregation rate. In this model $\theta^2 = 0$ and θ^3 is the rate of BP capture during the aggregation.

According to the mixture theory

$$\rho^1 = (1 - H)\rho_f, \quad \rho^2 = C\rho_s, \quad \rho^3 = (H - C)\rho_f, \tag{12}$$

where ρ_s, ρ_f are physical densities of RBC and BP.

Since the tube is thin ($d << L$), the radial and azimuthal motion can be neglected in comparison with vertical sedimentation and (11) can be written in 1d form.

Then substitution of (12) into the first Eq. (11) gives:

$$\frac{\partial(1 - H)}{\partial t} + \frac{\partial(1 - H)v^1}{\partial x} = -\theta;$$
$$\frac{\partial C}{\partial t} + \frac{\partial Cv^2}{\partial x} = 0; \tag{13}$$
$$\frac{\partial(H - C)}{\partial t} + \frac{\partial(H - C)v^3}{\partial x} = \theta,$$

where v^2 are vertical components of velocity vectors, $\theta = \theta^3/\rho_f$.

Summation of (13) gives the incompressibility condition for the mixture

$$\frac{\partial}{\partial x}\left[(1 - H)v^1 + Cv^2 + (H - C)v^3\right] = 0. \tag{14}$$

From the impermeability condition at the bottom of the tube

$$(1 - H)v^1 + Cv^2 + (H - C)v^3 = 0 \qquad (15)$$

and now from (14) one can obtain that (15) is valid at any cross section of the tube. According to the thermodynamic model of 3-phase thixotropic fluid [27, 28]

$$
\begin{aligned}
P^1_{lk} &= \left[(p + p^1_H)(1 - H) - \mu_1 e^1_{ii} \right] \delta_{lk} - 2\mu_2 e^1_{lk}, \\
P^2_{lk} &= \left[(p - p^1_s - p^2_s - p^3_s - p^3_H)C - \mu_3 \left(e^2_{ii} - \omega_{ii} \right) \right] \delta_{lk} - 2\mu_4 (e_{lk} - \omega_{lk}), \\
P^2_{lk} + \sigma_{lk} &= \left[(p - p^1_s - p^2_s - p^3_s - p^3_H)C - \mu_5 \omega_{ii} \right] \delta_{lk} - 2\mu_6 \omega_{lk}, \\
P^3_{lk} &= \left[(p - p^2_H - p^3_H)(H - C) - \mu_7 e^3_{ii} \right] \delta_{lk} - 2\mu_8 e^3_{lk}, \\
R^1_l &= F\left(v^2_l - v^1_l \right) - p \frac{\partial H}{\partial x_l} + p^1_s \frac{\partial C}{\partial x_l}, \\
R^3_l &= D\left(v^2_l - v^3_l \right) + (p - p^2_H) \frac{\partial (H - C)}{\partial x_l} + (p^3_s + p^3_H) \frac{\partial C}{\partial x_l},
\end{aligned}
\qquad (16)
$$

where μ_j, F, D are thermodynamic coefficients.

Substituting (16) into (14) and neglecting inertia, viscous stresses and assuming $p^\alpha_s = p^\alpha_H = 0$ $(\alpha = 1, 3)$, $\varepsilon_{ll} = 0$ one can obtain the following momentum equations

$$
\begin{aligned}
&-(1 - H) \frac{\partial p}{\partial x} + (1 - H)\rho_f G + F\left(v^2 - v^1 \right) = 0, \\
&-C \frac{\partial p}{\partial x} + \frac{\partial p^2_s C}{\partial x} + p^2_H \frac{\partial (H - C)}{\partial x} - F\left(v^2 - v^1 \right) - D\left(v^2 - v^3 \right) + C\rho_s G = 0, \quad (17) \\
&-(H - C) \frac{\partial (p - p^2_H)}{\partial x} + D\left(v^2 - v^3 \right) + (H - C)\rho_f G,
\end{aligned}
$$

where G(x) is external non-uniform force; in the gravity field G(x) = g.

From (15), (17) we have

$$
\begin{aligned}
v^1 &= -\frac{H(1 - H)}{F} \xi_1 + \frac{(H - C)^2}{D} \xi_2, \\
v^2 &= \frac{(1 - H)^2}{F} \xi_1 + \frac{(H - C)^2}{D} \xi_2, \\
v^3 &= \frac{(1 - H)^2}{F} \xi_1 + \frac{(H - C)(1 - H + C)}{D} \xi_2,
\end{aligned}
\qquad (18)
$$

where $\xi_1 = C(\rho_s - \rho_f)G + \frac{\partial}{\partial x} \left[p^2_s C + p^2_H (H - C) \right]$, $\xi_2 = \xi_1 - \frac{\partial p^2_H}{\partial x}$.

Now let us introduce an average volume of aggregates $w = C/N$ instead of N. Then from (18) one can derive the hyperbolic quasi-linear system of partial differential equations

$$
\frac{\partial w}{\partial t} + v^2 \frac{\partial w}{\partial x} = -\frac{\Gamma w^2}{C},
$$

$$
\frac{\partial H}{\partial t} + \gamma_{11}\frac{\partial H}{\partial x} + \gamma_{12}\frac{\partial C}{\partial x} + \gamma_{13}\frac{\partial w}{\partial x} = \theta, \qquad (19)
$$

$$
\frac{\partial C}{\partial t} + \gamma_{21}\frac{\partial H}{\partial x} + \gamma_{22}\frac{\partial C}{\partial x} + \gamma_{23}\frac{\partial w}{\partial x} = 0,
$$

where

$$
\gamma_{11} = v^1 - (1-H)\frac{\partial v^1}{\partial H}, \quad \gamma_{12} = -(1-H)\frac{\partial v^1}{\partial C}, \quad \gamma_{13} = -(1-H)\frac{\partial v^1}{\partial v},
$$

$$
\gamma_{21} = C\frac{\partial v^2}{\partial H}, \quad \gamma_{22} = v^2 + C\frac{\partial v^2}{\partial C}, \quad \gamma_{23} = C\frac{\partial v^2}{\partial w},
$$

$$
v^1 = \left[-\frac{H(1-H)}{F} + \frac{(H-C)^2}{D} \right] C(\rho_s - \rho_f)G, \quad v^2 = \left[\frac{(1-H)^2}{F} + \frac{(H-C)^2}{D} \right] C(\rho_s - \rho_f)G.
$$

The system (19) must be solved at the following initial and boundary conditions

$$
H(0,x) = H_0, \; C(0,x) = C_0, \; w(0,x) = w_0,
$$
$$
v^1(t,L) = 0, \quad v^2(t,L) = 0, \qquad (20)
$$

where H_0, C_0, w_0 are initial values that are supposed to be constant in the sample.

Solution of (19) and (20) can be found by the method of characteristics. The system (19) has three families of characteristics

$$
\left(\frac{dx}{dt}\right)_1 = v^2,
$$

$$
\left(\frac{dx}{dt}\right)_{2,3} = \frac{\gamma_{11} + \gamma_{22} \pm \sqrt{(\gamma_{11} - \gamma_{22})^2 + 4\gamma_{12}\gamma_{21}}}{2} \qquad (21)
$$

with corresponding conditions

$$
\frac{dw}{dt} = -\frac{\Gamma v^2}{C},
$$

$$
\left[\left(\frac{dx}{dt}\right)_{2,3} - \gamma_{22} \right] \left(\frac{dH}{dt} - \theta\right) + \gamma_{12}\frac{dC}{dt} = 0. \qquad (22)
$$

Let us introduce the following velocities of the solid and fluid phases

$$v_s = (1 - H)^2 C(\rho_s - \rho_f)G/F,$$
$$v_f = (H - C)^2 C(\rho_s - \rho_f)G/D, \tag{23}$$

then $v^1 = -Hv_s/(1 - H) + v_f$, $\quad v^2 = v_s + v_f$ and the expressions for the coefficients in (19) will be

$$\gamma_{11} = v_s + v_f + H\frac{\partial v_s}{\partial H} - (1 - H)\frac{\partial v_f}{\partial H}, \quad \gamma_{22} = v_s + v_f + C\frac{\partial v_s}{\partial C} + C\frac{\partial v_f}{\partial C},$$
$$\gamma_{12} = H\frac{\partial v_s}{\partial C} - (1 - H)\frac{\partial v_f}{\partial C}, \quad \gamma_{21} = C\frac{\partial v_s}{\partial H} + C\frac{\partial v_f}{\partial H}. \tag{24}$$

Substitution of (23) into (21) and (22) gives

$$\left(\frac{dx}{dt}\right)_1 = v_s + v_f,$$
$$\left(\frac{dx}{dt}\right)_{2,3} = v_s + v_f + \frac{1}{2}A\left(1 \pm \sqrt{1 - B/A^2}\right), \tag{25}$$

$$\frac{dw}{dt} = -\frac{\Gamma v^2}{C},$$
$$\frac{1}{2}\left[A_1(1 + \sqrt{1 - B/A^2}) - A_2(1 - \sqrt{1 - B/A^2})\right]\left(\frac{dH}{dt} - \theta\right) + \gamma_{12}\frac{dC}{dt} = 0,$$
$$\frac{1}{2}\left[A_1\left(1 - \sqrt{1 - B/A^2}\right) - A_2\left(1 + \sqrt{1 - B/A^2}\right)\right]\left(\frac{dH}{dt} - \theta\right) + \gamma_{12}\frac{dC}{dt} = 0, \tag{26}$$

where

$$A_1 = H\frac{\partial v_s}{\partial H} - (1 - H)\frac{\partial v_f}{\partial H}, \quad A_2 = C\frac{\partial v_s}{\partial C} + C\frac{\partial v_f}{\partial C}, \quad A = A_1 + A_2,$$
$$B = 4C\left(\frac{\partial v_s}{\partial H}\frac{\partial v_f}{\partial C} - \frac{\partial v_s}{\partial C}\frac{\partial v_f}{\partial H}\right).$$

Based on the dimension analysis, one may have

$$F = \mu w^{-2/3}\phi_1(H, C, \chi_i), \quad D = \mu v_0^{-2/3}\phi_2(H, C, \xi_i), \tag{27}$$

where ϕ_1, ϕ_2 are dimensionless functions that can be derived from experiments in the form

$$\phi_1 = \alpha H(1 - H)^{-\eta_1}, \quad \phi_2 = \beta C(1 - C/H)^{-\eta_2}$$

where α, β, η_1, η_2 are positive constants.

Finally, we have from (27)

$$F = \alpha\mu H(1-H)^{-\eta_1}(H/C)^{2/3}w^{-2/3}, \quad D = \beta\mu C(1-C/H)^{-\eta_2}w_0^{-2/3}. \quad (28)$$

Substitution of (28) into (18) gives the expressions for velocities

$$v_s = \frac{(1-H)^{2+\eta_1}(C/H)^{1/3}w^{2/3}(\rho_s-\rho_f)G}{\alpha\mu}, \quad v_f = \frac{(1-H)^2(1-C/H)^{\eta_2}w_0^{2/3}(\rho_s-\rho_f)G}{\beta\mu}.$$

Now one can obtain

$$
\begin{aligned}
\frac{\partial v_s}{\partial H} &= -\frac{1+(3\eta_1+5)H}{3H(1-H)}v_s, & \frac{\partial u_s}{\partial C} &= \frac{1}{3C}v_s \\
\frac{\partial v_f}{\partial H} &= \frac{2+\eta_2 C/H}{H-C}v_f, & \frac{\partial u_f}{\partial C} &= -\frac{2+\eta_2}{H-C}v_f
\end{aligned}
\quad (29)
$$

Accounting for (29), the coefficients A_1, A_2, A, B, γ_{12} in (26) will have the form

$$
\begin{aligned}
A_1 &= -\frac{1+(3\eta_1+5)H}{3(1-H)}v_s - \frac{(2+\eta_2 C/H)(1-H)}{H-C}v_f \\
A_2 &= \frac{1}{3}u_s - \frac{(2+\eta_2)C}{H-C}u_f \\
A &= -\frac{(\eta_1+2)H}{1-H}v_s - \left[\frac{2(1-H+C)+\eta_2 C/H}{H-C}\right]v_f \\
B &= \frac{4C}{H-C}\left[\frac{1+(3\eta_1+5)H}{3H(1-H)}(2+\eta_2) - \frac{(2+\eta_2 C/H)}{3C}\right]v_s v_f \\
\gamma_{12} &= \frac{H}{3C}v_s + \frac{(1-H)(2+\eta_2)}{H-C}v_f .
\end{aligned}
\quad (30)
$$

Now numerical computations on (25) and (26) can be carried out. At the beginning of the RBC sedimentation $v_f \ll v_s$ because $(H-C)/(1-H) \ll 1$. Then one can neglect the terms $\sim v_f/v_s$ and rewrite (25) and (26) in a simple form

$$\left(\frac{dx}{dt}\right)_1 = v_s, \quad \left(\frac{dx}{dt}\right)_2 = -\left[\frac{(\eta_2+2)H}{1-H}-1\right]v_s, \quad \left(\frac{dx}{dt}\right)_3 = v_s, \quad (31)$$

$$\frac{dw}{dt} = -\frac{\Gamma v^2}{C}, \quad \left[-\frac{1+(3\eta_1+5)H}{3(1-H)}\right]\left(\frac{dH}{dt}-\theta\right) + \frac{H}{3C}\frac{dC}{dt} = 0, \quad \frac{dH}{dt} - \frac{H}{C}\frac{dC}{dt} = \theta. \quad (32)$$

At the end of the test the RBC are no longer in a free fall and $u_s \ll u_f$ [32]. In this case one can neglect in (26) the terms $\sim u_s/u_f$ and assume $H \sim 1$, and have (26) and (27) as the following

$$\left(\frac{dx}{dt}\right)_1 = v_f, \quad \left(\frac{dx}{dt}\right)_2 = -\left[\frac{2(1-H+C)+\eta_2 C/H}{H-C}\right]v_f, \quad \left(\frac{dx}{dt}\right)_3 = v_f, \quad (33)$$

$$\frac{dw}{dt} = -\frac{\Gamma v^2}{C}, \quad -\left(\frac{dH}{dt} - \theta\right) + \left(\frac{2 + \eta_2}{2 + \eta_2 C/H}\right)\frac{dC}{dt} = 0,$$

$$C\left(\frac{dH}{dt} - \theta\right) + (1 - H)\frac{dC}{dt} = 0. \tag{34}$$

As one can see from (26), the first characteristics is always positive $((dx/dt)_1 > 0)$, third one is negative $((dx/dt)_3 < 0)$, while the second one $\left(\frac{dx}{dt}\right)_3 > 0$ when $H \in \left(0, (\eta_1 + 3)^{-3}\right)$ and $\left(\frac{dx}{dt}\right)_3 < 0$ when $H \in \left((\eta_1 + 3)^{-3}, 1\right)$.

Numerical computations on (26) and (27) have been carried out using the model parameters for the blood samples examined in the experiments (see Chap. 2)

$$H_0 = 30 \div 50\%, \ \mu_f = (1.1 \div 1.7) \cdot 10^{-3} \, \text{Pa} \cdot \text{s}, \ G = g,$$

$$\rho_f = 1030 \div 1080 \, \text{kg/m}^3, \ \rho_s = 1050 \div 1150 \, \text{kg/m}^3.$$

4 Results and Discussions

When the BP capture inside the aggregates can be neglected, from (26) and (27) one can obtain two characteristics; the positive one started from the top of the tube $x = 0$ and the negative one started from the bottom $x = L$ (Fig. 3). For simplicity, the vertical axes is located down along the gravity field (Fig. 3a) and the regions 1, 2, 3 (Fig. 3b) correspond to the zones of clear BP, settling aggregates and compact RBC network [32]. The two characteristics meet at $t = t_{max}$ when the ESR reach its maximal value (Fig. 3). At $t > t_{max}$ the compact RBC network influence the settling aggregates that reducing the ESR.

Based on (25)–(27), the following expression for the time t_{max} has been obtained

$$t_{max} = \frac{1}{kH_0}\left[\left(\frac{5kH_0 f(H_0)}{3[1 - \gamma(H_0)](1 - H_0)^2} + 1\right)^{0.6} - 1\right], \tag{35}$$

where $\gamma(H) = 1 + (H/v)/(\partial v/\partial H)$.

When the BP capture inside the aggregates is essential, three families of characteristics produce four regions by their cross sections (Fig. 4a, b). Depending whether the third characteristics is positive or negative, two different distributions can be seen. The first zone again corresponds to the clear BP region, while zone 2 is field with single RBC and small aggregates distributed in BP. Zone 3 corresponds to the aggregating RBCs and settling aggregates, while zone 4 is the compact zone composed by the resting aggregates with a small amount of BP. Depending on the model parameters the height h_2 of the zone 2 may be small ($h_2 \ll h_3$, Fig. 4a) or big ($h_2 \gg h_3$, Fig. 4b) at the same values h_1, h_4. In the experiments both cases have been observed. The zone 2 has been present and semi-transparent pink or red colored region. Most likely, the zone

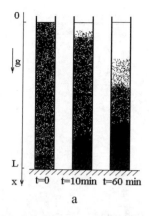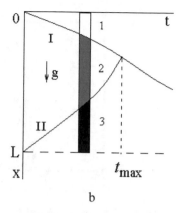

Fig. 3. Schematic representation of the sedimentation dynamics at different times (a) according to two families of characteristics (I, II) (b).

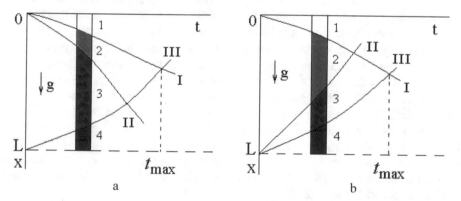

Fig. 4. Schematic representation of the sedimentation dynamics according to three characteristics (I, II, III) with positive (a) and negative (b) 2-nd characteristic.

2 correspond to the BP released at high vertical velocities from the aggregates due to instability and rapid compression of the RBC network in the zone 3 [32].

The software allowed numerical computations of the concentrations and velocities in each zone based on the method of characteristics has been developed. The mesh composed by three families of characteristics is depicted in Fig. 5. The convergence has been reached very fast starting with the division of the length $X \in [0, 1]$ into 20 segments. The location of the cross section point (t_{max}) is strongly determined by the model parameters.

Some numerical results are presented in Figs. 6 and 7a–c. Three types of sedimentation dynamics have been observed. At some combinations of the model parameters the ESR almost linear increases during the first 20–35 min of sedimentation and then sedimentation almost stopped (Figs. 6a and 7a). In other cases the ESR curves are S-shaped with initial slow ESR replaced by faster sedimentation, which is decelerated then till to the constant sedimentation velocity (Figs. 6c and 7c). When initial

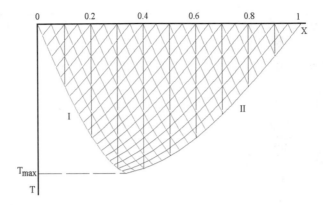

Fig. 5. Two families (I and II) of characteristics and determination of t_{max}.

acceleration was non-linear but without typical S-shape, the ESR curves were classified as intermediate ones (Figs. 6b and 7b). The main factors influenced the initial acceleration of sedimentation were initial RBC concentration H_0, BP viscosity μ_f, and RBC aggregation rate k. At low $H_0 \leq 35\%$ slow linear increase $h_1(t)$ has been observed (Fig. 6a). The 60 min values of h_1 corresponded to healthy organisms at $k = 10^5 - 10^6$, while $k = 10^7 - 10^9$ were proper to the h_1 values corresponded to pneumonia, moderate inflammations, and some other diseases [2, 3]. At low initial concentrations and elevated BP viscosity or/and density the intermediate cases have been observed (Fig. 6b). In this case the same influence of the aggregation rate has been revealed. When $H_0 \geq 45\%$ and ρ_f, μ_f are increased, the typical S-shaped ESR curves have been computed. Those combinations of the rheological parameters correspond to unclear BP filled with reactive proteins [5, 6] that is proper to severe inflammations, and cancer [2, 3].

Based on the computation data, the values t_{max} have been computed from the differential curves $h'(t)$ and on (35). It was shown, the characteristic values of the RBC aggregation rate k measured on the control group, were in the range $k \in [1.2 \cdot 10^5, 2.9 \cdot 10^7]$, while in the patients with allergy $k \in [9.3 \cdot 10^8, 7.9 \cdot 10^9]$ [24]. Similar computations on the ESR curves measured on the patients with different types of cancer [32] revealed the values $k > 10^{10}$. Since cancer leads to significant variations of thickness and structure of RBC hydration shells [20], the observed extremely elevated ESR values are connected with RBC surface changes by cancer disease, their interconnected electrical and mechanical properties. It was show, the RBC surface properties, structure and thickness of their hydration shells restore after successful X-ray and/or chemotherapy that is shown by their dielectric and temperature spectra [33]. Therefore, the ESR curves can be also used for control of the outcomes of the prescribed treatment and prognosis of disease. This important clinical application must be first tested on several groups of patients and healthy donors that will be done in out next study.

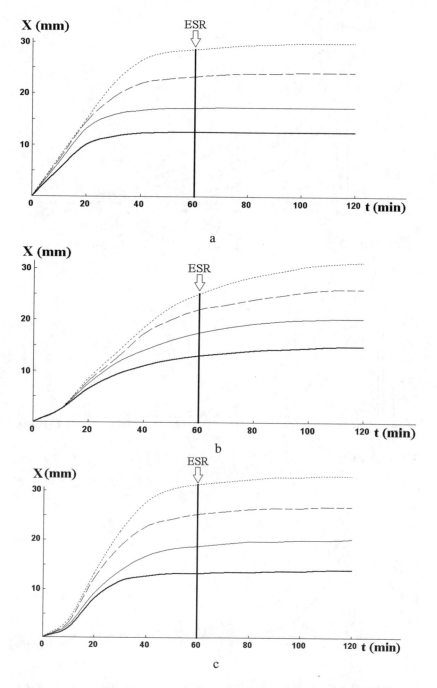

Fig. 6. ESR curves X(t) at $H_0 = 30\%, \mu = 5 \cdot 10^{-3} Pa \cdot s$ (a), $H_0 = 40\%, \mu = 4 \cdot 10^{-3} Pa \cdot s$ (b), $H_0 = 48\%, \mu = 3 \cdot 10^{-3} Pa \cdot s$ (c), at $k = 10^5$ (thick line), $k = 10^6$ (thin line), $k = 10^7$ (dashed line), $k = 10^8$ (dotted line).

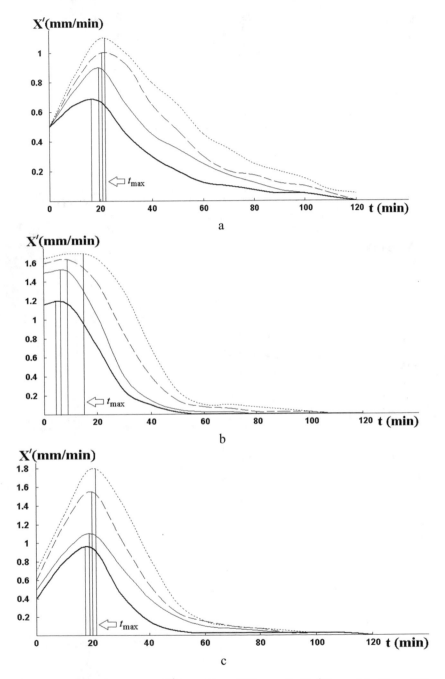

Fig. 7. Differential ESR curves X'(t) at $H_0 = 30\%, \mu = 5 \cdot 10^{-3} Pa \cdot s$ (a), $H_0 = 40\%, \mu = 4 \cdot 10^{-3} Pa \cdot s$ (b), $H_0 = 48\%, \mu = 3 \cdot 10^{-3} Pa \cdot s$ (c), at $k = 10^5$ (thick line), $k = 10^6$ (thin line), $k = 10^7$ (dashed line), $k = 10^8$ (dotted line).

5 Conclusions

The ESR is a very simple diagnostic test, which is widely used for non-specific diagnostics of the state of an organism, possible inflammations, ageing and mental disorders and some specific diseases. The ERS curves X(t) measured by computer assisted equipment carry more information on aggregation dynamics and specific interactions between RBC, BP proteins, products of the soft-tissue decay, apoptosis or new growth. The differential $h'(t)$ curves reveal stable time intervals t_{max} when $h'(t) \rightarrow$ max. A modification of the ESR test when a small portion of potential allergen (pollen, drug or food allergens) is added to the stabilized native blood is very helpful in revealing different food and drug allergy. It gives a new tool for fast and cheap quantitative estimation of the allergy and personal compatibility with prescribed medicine. An addition of the allergen can lead either to acceleration or to deceleration of the ESR in comparison with the control probe. Due to instability of the sedimentation process when occasional formation of a large aggregate at the top of the sedimentation tube could fall down faster and accelerate the process, the study of several (3–5) probes with the same allergen could be recommended. The decision can be made on the averaged value of t_{max} and averaged ESR curves.

References

1. Fung, Y.C.: Biomechanics. Circulation. Springer, New York (1997). https://doi.org/10.1007/978-1-4757-2696-1
2. Baskurt, O.K., Hardeman, M.R., Rampling, M.W.: Handbook of Hemorheology and Hemodynamics. IOS Press, Amsterdam (2007)
3. Olshaker, J.S., Jerrard, D.A.: The erythrocyte sedimentation rate. J. Emergency Med. **15**(6), 869–874 (1997)
4. Calderon, A.J., Wener, M.H.: Erythrocyte sedimentation rate and C-reactive protein. Hosp. Med. Clin. **193**, 313–337 (2012)
5. Litao, M.K., Kamat, D.: Erythrocyte sedimentation rate and C-reactive protein: how best to use them in clinical practice. Pediatr. Ann. **43**(10), 417–420 (2014)
6. Kainth, M.K., Gigliotti, F.: Simultaneous testing of erythrocyte sedimentation rate and C-reactive protein: increased expenditure without demonstrable benefit. J. Pediatr. **165**(3), 625–627 (2014)
7. Grzybowski, A., Sak, J.: Edmund Biernacki (1866–1911): discoverer of the erythrocyte sedimentation rate. On the 100th anniversary of his death. Clin. Dermatol. **29**(6), 697–703 (2011)
8. Daniels, L.M., Tosh, P.K., Fiala, J.A., et al.: Extremely elevated erythrocyte sedimentation rates: associations with patients' diagnoses, demographic characteristics, and comorbidities. Mayo Clinic Proc. **92**(11), 1636–1643 (2017)
9. Cengiz, O.K., Esmen, S.E., Varli, M., et al.: Markedly elevated erythrocyte sedimentation rate in older adults. How significant clinically? Eur. Geriatr. Med. **4**(1), 28–31 (2013)
10. Shteinshnaider, M., Almoznino-Sarafian, D., Tzur, I., et al.: Shortened erythrocyte sedimentation rate evaluation is applicable to hospitalised patients. Europ. J. Internal Med. **21**(3), 226–229 (2010)

11. Karlsson, H., Ahlborg, B., Dalman, Ch., Hemmingsson, T.: Association between erythrocyte sedimentation rate and IQ in Swedish males aged 18–20. Brain Behav. Immun. **24**(6), 868–873 (2010)
12. van den Broek, N.R., Letsky, E.A: Pregnancy and the erythrocyte sedimentation rate. Br. J. Obstet. Gynaecol. **108**(11), 1164–1167 (2001)
13. Choi, J.W., Pai, S.H.: Influences of hypercholesterolemia on red cell indices and erythrocyte sedimentation rate in elderly persons. Clin. Chim. Acta **341**(1–2), 117–121 (2004)
14. Ingelsson, E., Ärnlöv, J., Sundström, J., Lind, L.: Inflammation, as measured by the erythrocyte sedimentation rate, is an independent predictor for the development of heart failure. J. Amer. Coll. Cardiol. **45**(11), 1802–1806 (2005)
15. Rabjohn, L., Roberts, K., Troiano, M., Schoenhaus, H.: Diagnostic and prognostic value of erythrocyte sedimentation rate in contiguous osteomyelitis of the foot and ankle. J. Foot Ankle Surg. **46**(4), 230–237 (2007)
16. Glaser, R.: Biophysics: An Introduction. Springer, Berlin (2012)
17. Alison, J., Sheppard, R.: Dielectric properties of human blood at microwave frequencies. Phys. Med. Biol. **38**, 971–978 (1993)
18. Jaspard, F., Nadi, M., Rouane, A.: Dielectric properties of blood: an investigation of haematocrit dependence. Physiol. Meas. **24**, 137–147 (2003)
19. Lisin, R., Ginzburg, B.Z., Schlesinger, M., Feldman, Y.: Time domain dielectric spectroscopy study of human cells: I. Erythrocytes and ghosts. Biochim. Biophys. Acta **1280**(1), 34–40 (1996)
20. Batyuk, L.: Influence of cancer disease on dielectric characteristics of structural-functional state of erythrocyte membranes. ScienceRise Med. Sci. **7**(12), 11–17 (2015)
21. Bertoluzzo, S.M., Bollini, A., Rasia, M., Raynal, A.: Kinetic model for erythrocyte aggregation. Blood Cells Mol. Dis. **25**(2), 339–349 (1999)
22. Bell, G.I.: Models for the specific adhesion of cells to cells. Science **200**(1088), 618–627 (1978)
23. Chesnutt, J.K.W., Marshall, J.S.: Blood cell transport and aggregation using discrete ellipsoidal particles. Comput. Fluids **38**(6), 1782–1794 (2009)
24. Kizilova, N., Cherevko, V.: Mathematical modeling of particle aggregation and sedimentation in concentrated suspensions. In: Korzynski, M., Czwanka, J. (eds.) Mechanika w Medycynie, vol. 12, pp. 43–52. Rzeszow Univ. Press (2014)
25. Provata, A., Trohidou, K.N.: Spatial distribution and fractal properties of aggregating magnetic and non-magnetic particles. Fractals **6**(2), 219–230 (1998)
26. Neu, B., Miesleman, H.J.: Depletion-mediated red blood cell aggregation in polymer solutions. Biophys. J. **83**(5), 2482–2490 (2002)
27. Regirer, S.A.: On continual models of suspensions. Appl. Math. Mech. **42**(4), 679–688 (1978)
28. Regirer, S.A., Shadrina, N.H.: On models of tixotropic liquids. Appl. Math. Mech. **42**(5), 856–865 (1978)
29. Regirer, S.A.: Lectures on Biological Mechanics. Moscow University Press, Moscow (1980)
30. Kizilova, N.: Aggregation in magnetic field. In: Contemporary Problems of Biomechanics, vol. 9, pp. 118–135. Moscow University Press, Moscow (1994)
31. Chesnutt, J.K.W., Marshall, J.S.: Blood cell transport and aggregation using discrete ellipsoidal particles. Comput. Fluids **38**(5), 1782–1794 (2009)
32. Pribush, A., Meyerstein, D., Meyerstein, N.: The mechanism of erythrocyte sedimentation. Part 2: The global collapse of settling erythrocyte network. Colloids Surf. B Biointerfaces **75**(1), 224–229 (2010)
33. Batyuk, L., Kizilova, N.: Thermodynamic approach to dielectric parameters of human blood: application to early medical diagnostics of tumors. In: 14th Joint European Thermodynamics Conference, Book of Abstracts, Budapest (2017)

Modeling of Pulse Wave Propagation and Reflection Along Human Aorta

Natalya Kizilova[1]([⊠]), Helen Solovyova[2], and Jeremi Mizerski[3]

[1] Warsaw University of Technology,
Nowowiejska st., 24, 00-665 Warsaw, Poland
n.kizilova@gmail.com
[2] Kharkov National Polytechnic University,
Kirpichova st., 2, Kharkiv 61002, Ukraine
[3] Szpital Wojewódzki w Zamościu,
al. Jana Pawła II, 10, 22-400 Zamosc, Poland

Abstract. Pulse wave propagation, reflection and transmission along human aorta is studied on the 92-tube cadaveric model from aortic root to bifurcation. The branching coefficients, optimal coefficients by Murray, wave reflection coefficients by J. Lighthill have been computed and compared to the result computed on the 19-tube model of aorta derived from the 55-tube model of hyman systemic arterial tree by Westerhof. Variations in the local wave speed along the aorta have been computed on the model and compared to the continuous measurement data. It is shown the aorta is an optimal waveguide ensuring almost zero local wave reflections at the branches except for the aortic bifurcation, subclavian, carotid and kidney arteries. It is first shown that most of the branches have a negative wave reflection, which promotes blood acceleration and reduces the post-load on the heart due to the suction effect. The calculated values of the branching coefficients and pulse wave velocities correspond to the experimental measurements. The wave reflections at the kidney arteries depend on their individual geometry. The proposed approach can be used for preliminary estimation of the hemodynamic parameters caused by the wave propagation along individual aorta using the MRI study, and prediction of the risk of development of the cardiovascular diseases provided by abnormal hemodynamic.

Keywords: Pulse wave · Wave conductivity · Wave reflection
Medical diagnostics

1 Introduction

Aorta is the main vessel that distributes arterial blood to all parts and organs of the body through the systemic circulation. Periodic heart contractions generate small disturbances of pressure and velocity propagating from aorta through the arteries as waves with velocity $c = 5–10$ m/s which significantly exceeds the average speed of blood flow in the aorta $u = 0.5–0.8$ m/s. During the pulse wave propagation through the systemic arteries, the waves are repeatedly reflected at branching, narrowing (stenosis), enlargement (aneurysms), and at the regions of atherosclerotic or degenerative arterial walls. The physical parameters of the pulse waves and blood flow like pulse wave

© Springer Nature Switzerland AG 2019
K. Arkusz et al. (Eds.): BIOMECHANICS 2018, AISC 831, pp. 23–35, 2019.
https://doi.org/10.1007/978-3-319-97286-2_2

velocity (PWV) c, local linear velocity $u(t)$, fluctuations of blood pressure $p(t)$, vascular diameter $d(t)$ oscillations, and wall shear stresses (WSS) $\tau_w(t)$ are widely used for diagnostics of different cardiovascular pathology and diseases, the state of arteries and the blood supply to organs and tissues [1]. Therefore, the signals measured by ultrasound (US) equipment, magnetic resonance imaging (MRI) or direct methods [2] in the arteries are the result of the superposition of the incident, transmitted and numerous reflected waves. Analysis of the registered curves and biomechanical interpretation of their parameters are essential for advanced medical diagnostics [3].

Increased stiffness of the arterial walls due to atherosclerosis, persistent hypertension or other diseases leads to an increase in the PWV up to $c = 10$–15 m/s in the aorta and to $c = 15$–20 m/s in small arteries [4]. In this case, the reflected waves return to the aortic root much earlier than that of a healthy person, namely during systole time, not diastole, that leads to an increase in the amplitude of fluctuations in the blood pressure near the aortic valve, in the root and aortic arch that eventually leads to damage of the valve leaflets, the walls of aortic arch and its main branches. In young healthy individuals, PWV is markedly increased as the wave propagates along the aorta due to the gradual narrowing of the aorta. In elderly people PWV is high and weakly dependent on artery diameter due to degenerative age-related changes in arterial walls [5]. Significant changes in the physical characteristics of the vessel walls are also observed in diabetes. It has been shown that PWV is a statistically significant independent mortality risk index in groups of patients with type II diabetes and glucose tolerance [6].

For diagnostic purposes, contour, spectral and wavelet analyses of the measured $p(t), u(t), d(t)$ signals are used [1, 7]. An important physical parameter that determines the development of vascular pathologies is the mean WSS $\langle \tau_w(t) \rangle$ averaged over the period of heart contraction. The regions with lower $\langle \tau_w(t) \rangle$ values correlate with locations of atherosclerotic plaques [8]. The dependency p(u) gives information on the wave speed and compliancy of the vessel, while the phasic curves $u'(u), u''(u), d'(d), d''(d)$ where the stroke sign corresponds to derivative, are common in analyses of dynamical systems and give more diagnostic information [7]. During the last decades decomposition of the pressure and flow waveforms into their forward $p^+(t), u^+(t)$ and backward $p^-(t), u^-(t)$ running waves [9, 10] and analysis of the wave intensities $I^\pm(t) = p^\pm(t) \cdot u^\pm(t)$ (wave-intensity analysis, WIA) became an important diagnostic tool [11, 12]. The forward waves propagate downstream with blood flow and carry diagnostic information on the state of the heart and aorta, while the parameters of the backward running reflected waves depend on the state of blood circulation in the internal organs, muscles and tissues and, therefore, carry information on local microcirculation. The decomposition and WIA are based on mathematical models of pulse waves in arteries, taking into account the nonlinear properties of arterial wall, the non-Newtonian properties of blood, weakly and strongly nonlinear wave dynamics, which can lead to chaotic dynamics [13, 14].

Since individual features of the pulse wave propagation along aorta significantly affect the blood delivery to the main internal organs, their normal functioning, growth and development, elaboration of the patient-specific models of aorta and systemic arteries is essential for early diagnostics and long term risk estimation of development of the cardiovascular diseases provoked by abnormal hemodynamic. In this article the details of the pulse propagation and reflection along human aorta computed on five cadaveric models of aorta and its branches are presented.

2 Mathematical Models of Pulse Waves in Arteries

Linear 1d theory of waves in the fluid-filled compliant tubes was developed by J. Lighthill [15]. Numerical solution of 1d equation in an arbitrary tree of compliant tubes can be obtained by the Riemann method of characteristics. Linearized 2d theory of axisymmetric cylindrical pulse waves in the arteries was developed by J.R. Womersley in 1955–1957 [1].

The 1d theory of plane waves allows calculating the variations of pressure and velocity in an arbitrary section of the arterial vessel in the form of superposition of the forward and backward running waves [9–12]

$$
\begin{aligned}
p(t,x) &= p_0(f(t - x/c) + \Gamma f(t + x/c)), \\
u(t,x) &= \frac{p_0}{\rho c}(f(t - x/c) - \Gamma f(t + x/c)),
\end{aligned}
\tag{1}
$$

where p_0 is the amplitude of the wave at the inlet of the arterial segment, ρ is the blood density, c is the wave speed, $\Gamma = p^-/p^+$ is the coefficient of wave reflection at any heterogeneity like arterial branching, aneurism, stenosis or plaque, which is equal to the ratio of amplitudes p^+ and p^- of the incident and reflected waves.

The method of characteristics allows numerical calculations on the 1d model of quite complex arterial systems like 55-tube model of human systemic arterial tree from the aortic root to the main arteries of the upper and lower extremities [16].

If the input pressure wave generated by the heart is given in the form of a Fourier expansion $p(t, 0) = \sum_{k=0}^{\infty} p_k^0 e^{i\omega_k t}$, where p_k^0, ω_k are the amplitude and frequency of the k-th harmonic, then the expressions for pressure and velocity can be obtained in the form [1]

$$
\begin{aligned}
p_j(t, x_j) &= \sum_{k=0}^{\infty} p_{jk}^0 e^{i\omega_k t}\left(e^{-i\omega_k x_j/c_{jk}} + \Gamma_j e^{i\omega_k(x_j - 2L_j)/c_{jk}}\right) \\
Q_j(t, x_j) &= \sum_{k=0}^{\infty} Y_{jk}^0 p_{jk}^0 e^{i\omega_k t}\left(e^{-i\omega_k x_j/c_{jk}} - \Gamma_j e^{i\omega_k(x_j - 2L_j)/c_{jk}}\right)
\end{aligned}
\tag{2}
$$

where the index j refers to the tube number in the tree, Q_j is the average flow rate through the arterial cross section, $Y_{jk}^0 = (\pi d_j^2)/(4\rho c_{jk})$ is the characteristic wave admittance of the j-th tube for the k-th wave harmonic, $c_{jk} = \left(\frac{E_j h_j(1 - F_{jk})}{\rho d_j(1 - \sigma_j^2)} e^{I\theta_j}\right)^{1/2}$, $I = \sqrt{-1}$, $F_{jk} = 2J_1(\beta_{jk})/(\beta_{jk}J_0(\beta_{jk}))$, $\beta_{jk} = \alpha_{jk}(-1)^{3/4}$, $\alpha_{jk} = d_j\sqrt{\omega_k \rho/\mu_j}/2$ is the Womersley number, h_j, E_j, σ_j are the wall thickness, Young's modulus and Poisson ratio, μ_j is the blood viscosity, which depends on the vascular diameter d_j due to the non-Newtonian properties of blood in small arteries, J_0 and J_1 are the Bessel functions of the first kind of orders 0 and 1, respectively.

When the characteristic wave admittances of all segments of the arterial tree are computed, one can calculate the reflection coefficients of the waves at the branching of the parent vessel $(j = 0)$ into two daughter vessels $(j = 1, 2)$ by Lighthill's formula [15]:

$$\Gamma = \frac{Y_0^0 - Y_1^0 - Y_2^0}{Y_0^0 + Y_1^0 + Y_2^0}. \tag{3}$$

Usually, the wave admittances and reflection coefficients are calculated on the main harmonic $k = 1$, which transfers the largest portion of blood and contributes to blood supply. The results of the comparison of the computed local PWV, $p(t)$ and $u(t)$ curves showed a good conformity to the results of measurements in the aorta and large vessels [1, 3–5]. Solutions (2) are widely used for spectral analysis of pulse waves and impedance calculations of individual sections of the circulatory system [1].

The 2d theory equations are based on the Navier-Stokes equations for blood flow as a viscous incompressible fluid and the equations of the classical viscoelastic theory for the artery wall. Analytical solution of the system can be obtained in the form of a superposition of small disturbances (Fourier expansions) [1] for uniform or multilayer viscoelastic wall [14]. 2d models are very useful for analysis of instability of blood flow through collapsed arteries, as well as for study of high-frequency wall fluctuations and pathological noise generation over large arteries and veins [1].

3 Materials and Methods

The diameters d_j and lengths L_j of individual segments of the arteries from the aortic valve to the small extraorgan arteries were measured on five corpses in the course of postmortem examination [17]. In this state, the arteries are in the dilated state and their diameters are maximal that makes it possible to investigate the geometry of the whole systemic tree to the small arteries with diameters $d = 1$–1.5 mm that can not be done by non-invasive methods. At in vivo conditions the diameters of the same arteries could be smaller due to active tonic reactions of the smooth muscles in their walls. Individual topologic models of the vasculatures have been constructed as binary trees with some number of trifurcations (coronary, intercostal, lumbar and some others). Nodes and segments have been separately marked by two sets of numbers. Due to individual differences in the vascular geometry, the models had from 753 to 883 segments.

In this study the datasets corresponded to the aorta and its branches (Fig. 1) have been collected. The length along aorta from the aortic valve till the aortic bifurcation was $L_\Sigma = 60$–68 cm with 32 branches, which produced the reflected waves at each branch. The diameters measured in the middle section of each arterial segment and the lengths of the segments are presented in Table 1. The measured diameters have been interpolated to the inlet and outlet of each segment according to individual decline of

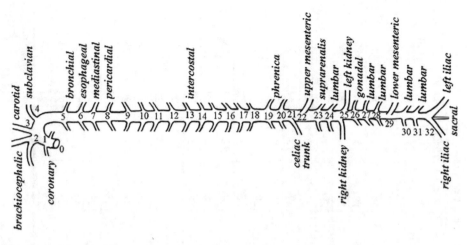

Fig. 1. A scheme of human aorta: left and right coronary arteries (1), brachiocephalic (2), left carotid (3), left subclavian (4), bronchial (5), esophageal (6), mediastinal (7), pericardial (8), 10 pairs of intercostal (9–18), superior (19) and inferior (20) phrenic arteries, celiac trunk (21), superior mesenteric (22), suprarenal (23), lumbar (24, 27, 28, 30, 31), renal (25), gonadal (26), inferior mesenteric (29), aortic bifurcation (32) with common iliac and middle sacral arteries.

the aorta $(d_0 - d_{21})/L_\Sigma$, and the diameters computed at the levels of bifurcations have been used for numerical calculations.

According to (3), the positive reflection ($\Gamma > 0$) means that the conductivity of the arterial system after the branching is lower than before it, so that in the latter cross section the amplitude of the pressure increases (damaging effect on the wall), and the blood flow decreases (ischemic action on the corresponding internal organs) [1]. The negative reflection ($\Gamma < 0$) leads to a decrease in the amplitude of the pressure oscillations and increased blood supply due to the suction effect [7]. The presence of negative reflection was shown in the coronary vasculature, as well as in the arterial tree of lungs and some other internal organs [7, 11, 13]. Here the peculiarities of the pulse wave propagation along the patient-specific model of aorta are studied, and wave speed and wave reflection coefficients are computed and compared to the measurement data.

4   Numerical Computations and Analyses

The following coefficients have been computed on the morphometric data:

- Branching coefficient $K = \frac{d_1^2 + d_2^2}{d_0^2}$ in the bifurcation or $K = \frac{d_1^2 + d_2^2 + d_3^2}{d_0^2}$ in the trifurcation. When $K > 1$, the total cross sectional area after the branching becomes bigger, which leads to decrease in the blood velocity, and vice versa;
- Optimality coefficient by Murray [7] in the bifurcation $M = \frac{d_1^3 + d_2^3}{d_0^3}$ or $M = \frac{d_1^3 + d_2^3 + d_3^3}{d_0^3}$ in the trifurcation. When $M \approx 1$, the branching produces minimal resistance for the steady flow;

Table 1. Diameters (d) and lengths (L) of the arterial segments in five humans h1, h2, h3, h4, h5 (height/age), males (m) and females (f).

N	Location	Name	h1(m) d(mm)	172/60 L(mm)	h2(m) d(mm)	177/58 L(mm)	h3(f) d(mm)	164/38 L(mm)	h4(f) d(mm)	155/43 L(mm)	h5(m) d(mm)	172/62 L(mm)
1	0–1	Ascending aorta I	33	3	33.3	3	32	3	29	3	35	3
2	1–2	Ascending aorta II	33	91	26.7	85	32	58	29	60	35	85
3	2–3	Aortic arch I	32	23	32	23	30	15	29	4	35	6
4	3–4	Aortic arch II	32	14	32	14	30	14.5	29	5	35	17
5	4–5	Thoracic aorta I	31	1	23.5	5	30	2	29	6	29	24
6	5–6	Thoracic aorta II	31	2	23	51	30	4	29	7	29	8.5
7	6–7	Thoracic aorta III	31	2.5	23	2	30	3	29	2	29	1.5
8	7–8	Thoracic aorta IV	31	21	22	22	30	23	29	12	29	9
9	8–9	Thoracic aorta V	31	18.5	22	17	30	19	29	8	29	4
10	9–10	Thoracic aorta VI	30.5	32	21.5	15	30	18	28	22	29	25
11	10–11	Thoracic aorta VII	30.5	32	21.5	15	29	18	28	22	29	25
12	11–12	Thoracic aorta VIII	30.5	32	21.5	15	29	18	28	22	29	25
13	12–13	Thoracic aorta IX	30	32	21.5	15	27	18	28	22	29	25
14	13–14	Thoracic aorta X	30	32	21	15	27	18	28	22	29	25
15	14–15	Thoracic aorta XI	29	32	21	15	25	18	28	22	29	25
16	15–16	Thoracic aorta XII	29	32	21	15	25	18	28	22	29	25
17	16–17	Thoracic aorta XIII	28	32	21	15	23	18	28	22	29	25
18	17–18	Thoracic aorta XIV	28	32	20.5	15	23	18	28	22	29	25
19	18–19	Thoracic aorta XV	26	18	19.5	17	20	6.5	23	4	26	14.7
20	19–20	Thoracic aorta XVI	25.7	20	19	18	20	8.5	23	4	26	27
21	20–21	Abdominal aorta I	24	9	18	8	20	6	22	11	26	13
22	21–22	Abdominal aorta II	23.6	33	17	44.5	19.5	12	22	13	25	34
23	22–23	Abdominal aorta III	23	1	19	8.5	19	10	20	4	24	6

(continued)

Table 1. (continued)

N	Location	Name	h1(m) d(mm)	172/60 L(mm)	h2(m) d(mm)	177/58 L(mm)	h3(f) d(mm)	164/38 L(mm)	h4(f) d(mm)	155/43 L(mm)	h5(m) d(mm)	172/62 L(mm)
24	23–24	Abdominal aorta IV	23	1	19	9.5	19	11	22	8	24	5
25	24–25	Abdominal aorta V	23	1	20	14	19	12	20.5	12	24	17
26	25–26	Abdominal aorta VI	22.5	20	20.5	14	2.9	11	20.5	8	23.5	15
27	26–27	Abdominal aorta VII	22.5	9	21	7	18.5	11	20	8	23	16
28	27–28	Abdominal aorta VIII	22	41	20	28	18	21	20	21	23	30
29	28–29	Abdominal aorta IX	22	18	20	8	18	13	20	18	23	16
30	29–30	Abdominal aorta X	22	23	19	22	17.5	11	19	12	21	16
31	30–31	Abdominal aorta XI	22	24	19	26	17	20	19	21	21	30
32	31–32	Abdominal aorta XII	22	23	20	24	17	19	19	17	21	28
33	1	Right coronary	3.1	37	3.5	38	6.3	26	3	48	2.7	35
34	1	Left coronary	3.9	29	4.9	29	6.5	24	3.8	11	2.9	15
35	2	Brachiocephalic trunk	19	42	11	36	13.5	35	11	36	16	31
36	3	Left common carotid	8.8	109	7.3	116	8.5	110	7.2	117	10.5	130
37	4	Left subclavian	12	64	13.7	44	8	26	7	29	10.5	35
38	5	Bronchial	1.8	36	1.4	35	1.2	30	1	23	1.3	28
39	6	Esophageal	2.4	28	1.8	27	1.6	25	0.8	19	1.2	26
40	7	Mediastinal	1.5	2	1.1	25	1.2	23	1.2	16	1.4	24
41	8	Pericardial	1.2	21	1	25.5	1.2	20.5	1.1	15	1.2	22
42	9	Intercostal I	3.2	23	2	22	2.9	12	2.5	9	2.5	14
43	10	Intercostal II	3.1	23	2.5	22	2.9	11	2.3	9	2.2	15
44	11	Intercostal III	3	23.5	2.7	22	2.9	10.5	2.3	8	2.2	15
45	12	Intercostal IV	3	23	2.5	22.5	2.9	11	2.3	8	2.2	14
46	13	Intercostal V	3	23	2.5	22	2.9	11.5	2.3	8	2.1	16

(continued)

Table 1. (continued)

N	Location	Name	h1(m) d(mm)	172/60 L(mm)	h2(m) d(mm)	177/58 L(mm)	h3(f) d(mm)	164/38 L(mm)	h4(f) d(mm)	155/43 L(mm)	h5(m) d(mm)	172/62 L(mm)
47	14	Intercostal VI	3.1	22.5	2.4	22	2.9	10.5	2.2	8	2.2	15
48	15	Intercostal VII	3	23	2.5	22	2.9	11	2.2	8	2.2	15
49	16	Intercostal VIII	3	23	2.5	22	2.9	11	2.3	8	2.2	16
50	17	Intercostal IX	3	23	2.4	21.5	2.9	1.5	2.3	7	2.1	16
51	18	Intercostal X	3	23	2.5	22	2.9	11	2.3	7	2.1	14
52	19	Phrenica superior	2	131	1.8	19	2.1	10	1.4	32	1.5	176
53	20	Phrenica inferior	2	131	1.8	18	2.2	9	1.5	36	1.6	155
54	21	Celiac trunk	8.9	11	7.5	11	8	17	8.7	14	10	15
55	22	Upper mesenteric	8	67	8.5	19	9.8	47	13	17	7	17
56	23	Suprarenalis	1	28	1.5	22	1.1	19	1	23	1.4	26
57	24	Lumbar I	2	46	1.8	158	3	56	1	23	2.2	38
58	25	Left kidney	6	41	4.3	53	8.5	87	5.6	37	4.4	68
59	25	Right kidney	5.4	28	4.2	69	8.5	45	6.7	47	4.1	54
60	26	Gonadal	1.5	310	1.3	340	1.8	300	1	223	2	250
61	27	Lumbar II	2	46	1.7	158	3	56	2.6	50	2.2	38
62	28	Lumbar III	2	46	1.7	158	3	56	2.6	50	2.2	38
63	29	Lower mesenteric	5.5	50	3.5	26	3.5	28.5	3	43	4	52
64	30	Lumbar IV	2	46	1.7	158	3	56	2.6	50	2.2	38
65	31	Left iliac	13.5	69	11	38	10	52	10	70	10	57
66	32	Lumbar V	2	46	2.1	158	3	56	2.6	50	2.2	38
67	32	Right iliac	13.3	67.5	11	53	10	52	8.3	53	12	53
68	32	Sacral	0.7	87	1.4	97	1.4	70	1.4	67	1.5	75

- Asymmetry coefficient $\chi = \min\{d_1, d_2\}/\max\{d_1, d_2\}$;
- Local PWV c at the first harmonics with $f = 1$ Hz;
- Characteristic wave admittance;
- Wave reflection coefficient Γ.

Distribution of the branching coefficient K averaged over the five sets of data is presented in Fig. 2, where L_Σ is the distance from aortic valve till the marked location along the aorta. For comparison the numerical results for original raw [16] and corrected [18] data are also presented. The prevailed number of branches except for the subclavian and aortic bifurcation has $K = 1 \pm 0.1$. Therefore, the linear velocity of the blood passed the branches remains almost constant. At the locations of the large brachiocephalic, renal, left subclavian and carotid arteries $K = 1.1$–1.3 and the linear velocity becomes slightly lower. Less pronounced effect with $K < 1$ is computed at the interior phrenic and superior mesenteric arteries. The highest deviation is observed at aortic bifurcation and at the level of renal arteries, that is known in physiology [1]. At the aortic bifurcation $K \approx 0.75$ and the blood flow into the iliac arteries is somehow accelerated. As it was shown on the canine circulatory system [19], at the aortic trifurcation $K \approx 1$ due to developed tail artery. In that way, some blood acceleration in humans might serve as a compensation of the noticeable wave reflection at the aortic bifurcation and backward blood flow. The numerical computations on nine aortic branches of the 55-tube model of human systemic arterial tree [16] (segments 5, 6, 7, 11, 14, 16, 19 in Fig. 1) revealed significant scatter of the branching coefficient around $K = 1$(rhombuses in Fig. 2). In that way, according to the raw data aorta is not an optimal waveguide with wave reflections and energy loss. The raw data on diameters were corrected in accordance to the hypothesis formulated on aorta as an optimal waveguide with zero reflections at the branches due to their well-matching [18]. The branching coefficients for the corrected diameters are marked by triangles in Fig. 2.

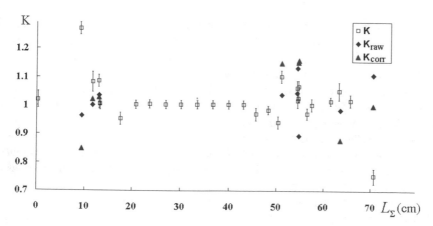

Fig. 2. Distributions of K along the aorta: $\square$ – computation data on 92-tube model on 5 datasets, $\Diamond$ - computations on [16], $\triangle$ - computations on[18].

Most of the smaller branches possess the Murray optimality coefficient $M \approx 1$ (Fig. 3) that is optimal for the steady flow, which corresponds to zero harmonics of the Fourier expansion of the pressure and flow waves. This harmonics transfer the most of the stroke volume of blood pumped by the left ventricle [20]. The larger branches of the aortic root, kidney and mesenteric arteries are close to the optimal values but their coefficient M exhibited some small scatter around the optimal value that might be normal for the wave flow [21]. Aortic bifurcation is the worst of all the branches in the meaning of high wave reflection. It was the scatter ± 0.1 around $M = 1$ corresponds to the branches in which the energy loss is only 5% bigger than the optimal ones [22]. The same computations on the raw [16] and corrected [18] data revealed similar distribution with $M = 1 \pm 0.15$ in 6 branches of 9 but with higher scatter than in the 92-tube model of aorta (Fig. 3).

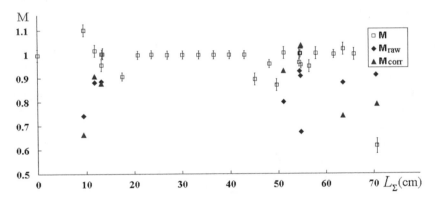

Fig. 3. Distributions of M along the aorta: □ – computation data on 92-tube model on 5 datasets, ◇ - computations on [16], ∆ - computations on [18].

The computed values of PWV along the aorta are given in Fig. 4. The obtained values are in a good agreement with other computational results and experimental in vivo measurements [16, 18]. In the larger elastic vessels c = 5–6 m/s that corresponds to healthy arteries. Monotonous increase in PWV along the aorta is determined by its narrowing, which becomes noticeable below the diaphragm ($L_\Sigma \sim 50$ cm in Fig. 2, 3, 4 and 5). Small decrease in PWV at the aortic bifurcation is defined by some increase in its lumen ($L_\Sigma \sim 72$ cm) [1].

The wave reflection coefficients computed along the aorta are presented in Fig. 5. The values of Γ are close to zero at almost all of the branches except for ones marked by arrows. At the most branches $\Gamma \in [-0.2, 0]$ and geometry of the aorta provides not only very small reflections due to matching of the vessels in each branch but slight suction effect. The latter promotes the blood inflow with wave components and decreases the hydrodynamic load on the left ventricle. The strongest negative reflections are proper to the left carotid and subclavian, superior and inferior mesenteric arteries (marked by 5–8 in Fig. 5). As one can see from Fig. 2, these locations possess $K > 1$. Positive reflections with quite small reflection coefficients $\Gamma < 0.05$ (i.e. with

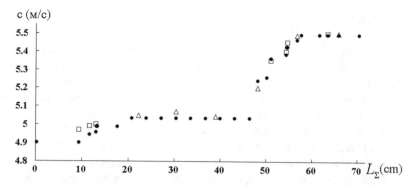

Fig. 4. Pulse wave speed along the aorta: ● - averaged computation data on 92-tube model, ☐ – computations on [16], Δ - in vivo measurements in 6 locations along aorta[18].

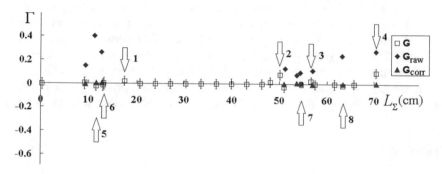

Fig. 5. Distribution of the wave reflection coefficient Γ along the aorta: : ☐ – computation data on 92-tube model on 5 datasets, ◇ - computations on [16], Δ - computations on[18].The arrows mark the beginning (1) and end (2) of the long segment with intercostal arteries, locations of the renal arteries (3), aortic bifurcation (4), left common carotid (5) and subclavian (6), superior (7) and inferior mesenteric (8) arteries.

energy loss <5%) are proper to bronchial, renal arteries and aortic bifurcation. It is well known, the wave reflections at aortic bifurcations could increase local pressure oscillations and lead to gradual damage of arterial wall and development of aneurism [1, 20]. High wave reflections at the renal arteries could also lead to gradual development of stenosis of one or both renal arteries [20, 23].

The reflection coefficients computed on raw data [16] are much higher $\Gamma \in [0, 0.4]$ (Fig. 5) and corresponds to poor-matched waveguide. That is why in [18] the diameters of the vessels have been corrected to have $\Gamma = 0$ at all the branches in agreement to the hypothesis on aorta as an optimal waveguide without wave reflections. According to our 92-tube model of aorta, most of its branches have small negative reflection coefficients and, thus, produce the suction effect for the blood flow into the downstream vasculature.

5   Conclusions

Based on numerical computations of the biomechanical parameters of steady (branching coefficient, Murray's coefficient) and wave (wave speed and wave reflection coefficient) flow on five cadaveric datasets, it was shown that the aorta is an optimal waveguide that provides steady blood flow (zero Fourier component) without significant decrease in the linear velocity ($K \sim 1$) and with minimal energy expenses ($M \sim 1$). Wave reflections along the aorta are close to zero due to well-matched branches ($\Gamma \sim 0$) with negative reflections when the reflected wave moves downstream and increases blood supply of the vasculature and inner organs with corresponding harmonics.

Calculated values of the wave speeds along aorta are in good correspondence with in vivo measured values [1, 18, 23]. The developed approach to aorta as a vessel directing blood flow and distributing it between the inner organs and muscles through aortic branches depending on its biomechanical parameters, may be used for advanced medical diagnostics. For this purpose, the dataset on individual geometry of aorta and its visible branches can be obtained with ultrasound or MRI technique. Modern MRI systems allow fast scanning of aortic tree, calculations of the corresponding lengths and diameters. Then the parameters k, M, c, Γ can be easily computed and analyzed. Mathematical analysis can determine the locations of low branching coefficients and high reflection coefficients due to individual geometry and predict possible development of the wall damage, development of stenosis or aneurism due to wave reflections.

References

1. Nichols, W., O'Rourke, M.: McDonald's Blood Flow in Arteries. Theoretical, Experimental and Clinical Principles. Hodder Arnold – Oxford University Press, New York (2005)
2. Grotenhuis, H.B., Westenberg, J.J.M., Steendijk, P., et al.: Validation and reproducibility of aortic pulse wave velocity as assessed with velocity-encoded MRI. J. Magn. Res. Imag. **30**(3), 521–526 (2009)
3. Latham, R.D., Westerhof, N., Sipkema, P., et al.: Regional wave travel and reflections along the human aorta: a study with six simultaneous micromanometric pressures. Circulation **72**(6), 1257–1269 (1985)
4. Rogers, W.J., Hu, Y.-L., Coast, D., et al.: Age-associated changes in regional aortic pulse wave velocity. J. Am. Coll. Cardiol. **38**(4), 1123–1129 (2001)
5. O'Rourke, M.F., Blazek, J.V., Morreels, C., Krovetz, L.J.: Pressure wave transmission along the human aorta. Circul. Res. **23**(10), 567–579 (1968)
6. Cruickshank, K., Riste, L., Anderson, S.G., et al.: Aortic pulse-wave velocity and its relationship to mortality in diabetes and glucose intolerance. Circulation **106**(4), 2085–2090 (2002)
7. Kizilova, N.: Novel aspects and perspectives of the theory of pulse waves in arteries. In: Chernyï, G.G., Regirer, S.A. (eds.) Contemporary Problems of Biomechanics, vol. 11, pp. 44–63. Moscow University Press, Moscow (2006)
8. Caro, C.G., Fitz-Gerald, J.M., Schroter, R.C.: Atheroma and arterial wall shear: observations, correlation and proposal of a shear dependent mass transfer mechanism for atherogenesis. Proc. Roy. Soc. Lond. Ser. B **177**(1), 109–159 (1971)

9. Westerhof, N., Sipkema, P., Bos, C.G.V., Elzinga, G.: Forward and backward waves in the arterial system. Cardiovasc. Res. **6**(4), 648–656 (1972)
10. Parker, K.H., Jones, J.H.: Forward and backward running waves in arteries: analysis using the method of characteristics. ASME J. Mech. Eng. **112**(2), 322–326 (1990)
11. Sun, Y.-H., Anderson, T.J., Parker, K.H., Tyberg, J.V.: Wave-intensity analysis: a new approach to coronary hemodynamics. J. Appl. Physiol. **89**(10), 1636–1644 (2000)
12. Li, Y., Parker, K.H., Khir, A.W.: Using wave intensity analysis to determine local reflection coefficient in flexible tubes. J. Biomech. **49**(6), 2709–2717 (2016)
13. Kizilova, N.: Pulse wave reflections in branching arterial networks and pulse diagnosis methods. J. Chin. Inst. Eng. **26**(6), 869–880 (2003)
14. Kizilova, N.: Blood flow in arteries: regular and chaotic dynamics. In:Dynamical systems. Applications. Awrejcewicz, J., Kazmierczak, M., Olejnik, P., Mrozowski, K. (eds). Lodz Politechnical University Press, Poland. 69–80 (2013)
15. Lighthill, M.J.: Waves in Fluids. Cambridge University Press, Cambridge (1978)
16. Westerhof, N., Bosman, F., de Vries, C.J., Noordegraaf, A.: Analog studies of the human systemic arterial tree. J. Biomech. **2**(1), 121–143 (1969)
17. Zenin, O., Gusak, V., Kirjakulov, G.: Human Arterial System in Numbers and Formulas. Donbass Pub., Ukraine (2002)
18. Alastruey, J., Khir, A.W., Matthys, K.S., et al.: Pulse wave propagation in a model of human arterial network: assessment of 1-D viscoelastic simulations against in vitro measurements. J. Biomech. **44**(5), 2250–2258 (2011)
19. Kizilova, N., Philippova, H., Zenin, O.: A realistic model of human arterial system: blood flow distribution, pulse wave propagation and modeling of pathology. In: Korzynski, M., Cwanka, J. (eds.) Mechanics in Medicine, vol. 10, pp. 103–118. Rzeszow University Press, Rzeszow (2010)
20. Folkov, B., Nil, E.: Circulation. London University Press, London (1971)
21. Li, Y., Parker, K.H., Khir, A.W.: Using wave intensity analysis to determine local reflection coefficient in flexible tubes. J. Biomech. **49**(5), 2709–2717 (2016)
22. Zamir, M., Bigelov, D.C.: Cost of depature from optimality in arterial branching. J. Theor. Biol. **109**(3), 401–409 (1984)
23. Milnor, W.R.: Hemodynamics. Williams & Wilkins, Baltimore (1989)

The Influence of Woman's Mastectomy on Breathing Kinematics

Frantisek Lopot[1], David Rawnik[2], Klara Koudelkova[1],
Petr Kubovy[1], and Petr Stastny[1(✉)]

[1] Faculty of Physical Education and Sport, Charles University,
Jose Martiho 31, 162 52 Prague, Czech Republic
flopot@seznam.cz, stastny@ftvs.cuni.cz
[2] University of Primorska, Lublan, Slovenia

Abstract. The aim of this study is to objectively determine the effect of total mastectomy on the extent of breathing movements of the thoracic and abdominal wall in the women (n = 6) who underwent this operation, compared to the breath movements of healthy women. Another aim is to find out whether there has been a symmetrical disturbance of the extent of movement between the operated and unoperated side of the chest and abdomen during breathing.

The 3D optoelectronic kinematic analyzer (Qualisys) has been used to measure the range of breathing movements. The chest mobility was measured in a calm and deep breathing. Further data on the patient's condition were obtained through a kinesiological analysis focused on the upper half of the body.

The mastectomy affects the extent of breathing movements in women who have undergone this operation, by generally reducing the range of breathing movements on the operative side of the fuselage. For this reason, the symmetry of the range of movements (ROM) between the operated and unoperated sides during breathing is also impaired. The most striking asymmetries are present approximately on the level of 5th ribs, where the postoperative scar is also the most common. These findings have been confirmed in both quiet and deep breathing, demonstrating our claim to the need for scar care.

Keywords: Respiratory movements · Total mastectomy · Breast cancer
Scar · Qualisys

1 Introduction

Total mastectomy along with axillary dissection (axillary dissection, sentinel node biopsy) is an extensive surgical procedure affecting soft tissues on the anterolateral side of the chest. As a result of surgery, the postoperative wound healing with a large scar. Due to the presence of the scar, numerous complications arise, such as myofascial pain and movement restriction. Extensive scars alter the mechanical behavior of the skin due to a natural tissue morphology disorder - the texture, the cleavage line described by Langer, Borges, Kraisel [1].

The scar differs from the intact tissue by a different arrangement of collagen fibers. Changing the fiber layout suggests a change in properties, especially a change in

© Springer Nature Switzerland AG 2019
K. Arkusz et al. (Eds.): BIOMECHANICS 2018, AISC 831, pp. 36–44, 2019.
https://doi.org/10.1007/978-3-319-97286-2_3

elasticity, with a loss of elasticity of up to 40% even in well-healed relatively large scars. Another difference is that scarred skin does not show a strict layered layout. The skin is not sutured in the surgical procedure. Sections are fastened separately, especially the skin, but the skin can't be sewn in isolation. The mutually shifting of the layers may subsequently exhibit other properties than the skin intact. Complications can also be a non-functional skin-to-subcutaneous tissue that limits the movement of tissues. In addition, there is a change in the extension of the strain along the line of the scar compared to the situation in the healthy tissue, the increase of the deformation of the skin in the remaining part and the formation of the significantly greater deformation force required to cause this deformation. This puts greater demands on the muscles involved and leads to overloading, pain, and long-term effects, as well as to the conversion of an optimal movement strategy that can contribute to undesirable structural changes in skeletal, muscle and lymphatic tissue [2].

Breathing movements in the chest area are likely to be affected by the scare healing. These movements are repeated cyclically, so even the slight increase in strain force that the muscles have to overcome can eventually lead to overloading. The subsequent respiratory disorder is usually associated with a movement disorder in a particular segment or sector of the spine and chest. This leads to the development of muscle imbalances and subsequent vertebral algic syndrome [3].

Because the presence of large scare can effect breathing functions, the objective of this study is to determine the effect of total mastectomy on the extent of breathing movements of the thoracic and abdominal wall in women who underwent this operation, compared to the breath movements of healthy women. Furthermore, we wanted to find out whether there is asymmetry of the range of motion between the operated and unoperated side of the chest and abdomen during breathing.

2   Methods

2.1   Participants

The research was attended by six women after total breast mastectomy, who underwent surgery between three months to 5 years. Our demands were to stop oncological treatment and not to resume breast reconstruction. In the sample, three women were on left-sided total mastectomy with axillary dissection, two women on right-sided total mastectomy with axilla dissection, and one left-sided woman without axillary performance. Four patients were complemented with chemotherapy with radiotherapy and two patients with radiotherapy. The average age of respondents is 64.5 years (range 58–79 years). The main characteristics of respondents are given in Table 1.

The control group consists of six healthy women who have not undergone any thoracic or abdominal surgery nor are they being treated with any serious illness. The mean age of women in the control group is 72.6 years. All the characteristics of respondents are given in Table 2.

The study was approved by the ethics committee of Charles university, Faculty of Physical Education and Sport, each respondent signed informed consent which is in accordance with the Helsinki Declaration.

Table 1. Basic characteristics of a group of women after total mastectomy.

Participant	1	2	3	4	5	6
Age (years)	58	68	79	66	58	58
Height (cm)	171	166	165	170	169	175
Weight (kg)	79	69	56	90	65	80
Time after operation (years)	5	19	25	22	6	5
Side of operation	Left	Right	Left	Left	Right	Left

Table 2. Basic characteristics of a group of women from the control group.

Participant	1	2	3	4	5	6
Age (years)	80	77	73	66	67	77
Height (cm)	160	152	162	161	167	160
Weight (kg)	72	54	81	85	72	71

2.2 Measurement

The Qualisys optoelectronic kinematic analyzer and Qualisys Track Manager to measure the range of breathing movements of the hull was used. Eight high-frequency cameras were used, with four cameras placed in the front of the room and four at the rear. Recording was sampled at 250 Hz. Before each set of measurements a calibration was performed to ensure the correct operation of the system and to achieve the maximum possible measurement accuracy. The measurement error for this optical system has been found to have absolute error 1.6 mm [4] or less [5, 6], which had a marginal effect of measured values.

For each respondent, 24 passive markers were used, 12 from the anterior side of the thorax and 12 from the dorsal side of the thorax. The markers were placed on each side in three vertical and four horizontal rows. The horizontal series (lines) were determined as follows: 1. Sternoclavicular articulation, 2. Second sterno-costal articulation, 3. Center of distance between the second and fourth horizontal series, 4. Submammary groove. The anterior vertical rows were determined by the right medio-clavicular line, the anterior median line, the left medio-clavicular line. The posterior vertical lines were defined by the left scapular line, the posterior midline, and the right scapular line (red markers in Fig. 1).

2.3 Procedures

Measurement of breath movements took place in two parts. The respondent had 15 min to calm her breath, then she was placed in a standardized position in her standing and her task was to breathe calm, smooth and naturally for one minute. Followed by deep breathing, the respondent was placed in a standardized sitting position, and after triggering the recording, she had three smooth breaths and exhalations. The recording time was set to 30 s. Overall, each probe received data from 1 min and 30 s, which subsequently served to calculate the key parameters on the thorax.

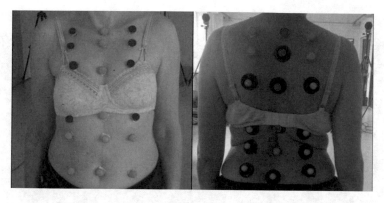

Fig. 1. Placement of markers on participants thorax.

2.4 Analyses

During the evaluation, individual markers were first manually identified and the course of their trajectories cleared. Subsequently, the distances between the individual markers placed on the thorax were chosen. Two distances were measured in each horizontal row, one tracking the left movement range and the second on the right side. The front right marker formed a pair with a rear middle marker, and the front left marker formed a pair with the same back middle marker (the red markers highlighted in Fig. 1). The distance between these markers was measured in all horizontal rows, so overall, eight distances were tracked.

The Qualisys Track Manager displays these distances by means of a graph (distance/time) using the resulting vector from the components x, y, z, where the change in distance between the markers for the inspiration and expiration is evident.

Custom data analysis was then performed in Microsoft Excel, where a sheet was created specifically for quiet and deep breathing. The minimum values, i.e., the breath, and the maximum values, i.e. the exhalation, were manually entered into the table in this program. After subtracting the individual values, the results were expressed as the movement of the thoracic wall in the individual breathing cycles, breath deflections. The first exhalation position was subtracted from the first inspirational position, then the second exhalation position was subtracted from the first inspirational position, then the second exhalation position was subtracted from the second exhalation position, etc., with the same procedure. Thus, the values were subtracted in each row to the left and to the right. The same method of calculation was used to process data in a quiet and deep breathing. The expiration position has always been subtracted from the inspiration to obtain only positive values. Figure 2 schematically shows the magnitude of breathing excursions from breath-inspiration (yellow arrow) and from breath-exhale (blue arrow).

With deeper breathing, when the recording lasted for 30 s, 3 breath cycles were observed. From the record of quiet breathing, which lasted for one minute, the first ten breathing cycles were used to measure breath deflections. From these values, the arithmetic average was finally calculated, resulting in the average breathing rate in one segment on one side - left or right.

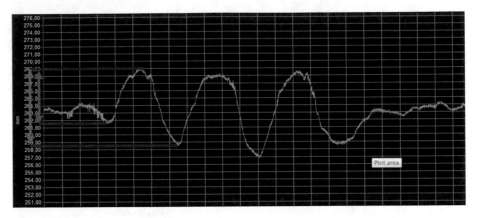

Fig. 2. Magnitude of breathing excursions from breath-inspiration (yellow arrow) and from breath-inspiration exhale (blue arrow).

The final parameter we observed was the ratio of the arithmetic mean breathing offsets operated to the unoperated side of the research group and the left to right ratio of the control group that was calculated for each horizontal row. This ratio gave us information about the difference in the extent of movement of the operated and unoperated side of the chest. Symmetry is a value of 1, numbers greater than 1 means higher use of the left/right side, numbers less than 1, on the other hand, higher use of the operative/right side of the chest.

2.5   Statistic

The data were processed using Statistica software and presented as means and standard deviations. Shapiro-Wilk, Levene and Mauchly's tests were used in order to verify the normality, homogeneity and sphericity of the sample's data variances, respectively. The effect size and mean percentage differences were used to estimate the magnitude of differences between mastectomy and healthy participants, where effect sizes were defined as large for $d > 0.8$, as moderate for between 0.8 and 0.5, and as small for <0.5 [7]. Differences of right and left kinematics ratios between mastectomy and healthy participants were analyzed using repeated measures ANOVA. The statistical significance was set at $p < 0.05$.

3   Results

The normality, homogeneity and sphericity were not disrupted in any selected group. The large effect size has been found between groups in thorax rows 1 to 4, only second line showed moderate effect during deep breathing (Table 3). The ANOVA showed differences between mastectomy and healthy group during in all four recorded lines on trunk in mild breathing ($F_{1,10} = 9.8$, $p = .010$) and deep breathing ($F_{1,10} = 20.4$, $p = .001$), see Figs. 3 and 4.

Table 3. Basic statistic of side to side thorax movement ration at different thorax rows.

	Mild breath				Deep breath			
Rows	Mastectomy	Healthy	Cohen d	Mean Δ	Mastectomy	Healthy	Cohen d	Mean Δ
	Mean ± SD	Mean ± SD		%	Mean ± SD	Mean ± SD		%
Line 1	0.88 ± 0.088	1.00 ± 0.099	1.27	11.0	0.84 ± 0.061	0.96 ± 0.85	1.58	14.0
Line 2	0.95 ± 0.072	1.03 ± 0.086	1.00	7.3	0.93 ± 0.010	1.00 ± 0.087	0.75	8.3
Line 3	0.84 ± 0.228	1.03 ± 0.078	1.11	18.6	0.85 ± 0.11	1.01 ± 0.020	2.06	14.5
Line 4	0.87 ± 0.155	0.98 ± 0.074	0.91	10.9	0.92 ± 0.104	1.02 ± 0.063	1.16	29.6

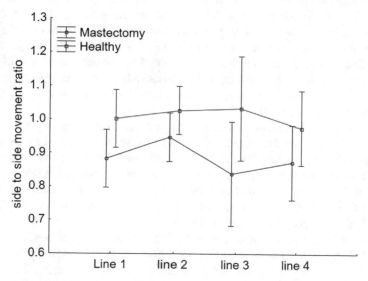

Fig. 3. Differences in mild breathing between participants after mastectomy and control group.

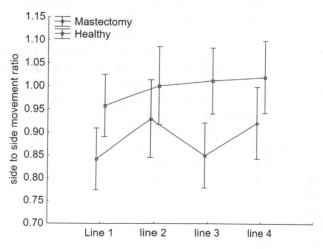

Fig. 4. Differences in deep breathing between participants after mastectomy and control group.

4 Discussion

Total mastectomy is a radical surgery that significantly affects the harmony of the human body, whether psychologically or physically. We have been dealing with the scar issue after this surgery and its influence on the motion apparatus with a focus on respiratory movements. When comparing the control and research group values, we found that a control group showed homogeneous results in the thoracic movement series. Symmetry of movements is very noticeable in quiet and deep breathing, compared to the large variance of values in the research group. Various ranges of observed values are apparent across the ranges. In the research group, the 2nd line is the most symmetrical, on the contrary, the 3rd row shows the greatest variance of breath movements between the sides in deep and quiet breathing (Figs. 3 and 4). We suppose it is precisely the presence of a scar that is the most devastating in this area. The smallest difference in the second series is explained by the stability of this area, which is ensured by relatively rigid chest bundle joints as well as musculature, which is in this area. The greater fixation and stability of the upper ribs in comparison with the lower ribs is reported by Kenyon [8].

Studies that would deal with a similar topic (total mastectomy on chest mobility during breathing) are scarce. Therefore, there is no data to confirm or refute our claim that total mastectomy reduces the mobility of the chest on the operative side. In our study, left-handed patients generally have a lower range of breathing movements on the left side of the thorax, with mild and deep breathing. For fair-operated women, this phenomenon is slightly less pronounced, but still in most lines there is a reduction in the extent of movement on the operated side.

We can at least mention studies that confirm the symmetry of breath movements between left and right in healthy people. De Groote et al. [9] measured the chest and abdominal movement during breathing in healthy respondents using 3D kinematic analysis, and did not mention that there was a difference in left and right movement among healthy respondents. The results of the Kaneko et Horie study [10] do not indicate that the difference between left and right side movements in the chest and abdomen and Ragnarsdóttir [11] measured the symmetry of these movements. These results therefore coincide with our findings in the control group.

A similar study, which examined the effect of thoracic surgery on breathing movements, is a study by Kristjánsdóttir [12], investigating the effect of cardiac surgery performed by media sternotomy. The measurements were performed before surgery, three months after surgery and one year after surgery. They found that three months after the operation, reduced upper chest mobility and increased abdominal wall movement were present compared to preoperative measurements. In addition, patients had a different range of movement between the left and right sides of the thoracic and abdominal wall after three months of surgery, a significant reduction in the left-hand movement range. Prior to surgery, these movements were symmetrical. After one year of surgery, there was still a reduction in movement in the upper chest area, especially on the left. Although this is a completely different operation, we can say that this study at least partially confirms our statement that chest surgery changes the breathing movements of the thorax.

Studies on postoperative complications after breast surgery most often follow the effect of the ROM in the shoulder joint [13], spinal kinematics [14], quality of life and the psychological state of the patient [15, 16]. Scar issue is often neglected in these studies, even though the need for scar care is widespread [17]. It is the scar that often causes pain, reduces ROM in the shoulder cluster, changes the kinematics of the spine and posture, the patient is mentally burdened and can promote the formation of lymphedema [2, 18].

5   Conclusion

The scar after total mastectomy affects the brainstem stereotype in women who have undergone the surgery. Generally, the range of breathing movements on the operating side of the thorax is reduced. These findings should be taken into account in postoperative rehabilitation, which should be started with early scar care.

References

1. Cerda, E.: Mechanics of scars. J. Biomech. **38**, 1598–1603 (2005)
2. Vránová, H., Zeman, J., Čech, Z., Otáhal, S.: Identification of viscoelastic parameters of skin with a scar in vivo, influence of soft tissue technique on changes of skin parameters. J. Bodyw. Mov. Therap. **13**, 344–349 (2009)
3. Véle, F.K.: Přehled kineziologie a patokineziologie pro diagnostiku a terapii poruch pohybové soustavy. 2. vyd. Triton, Praha (2006). 375 str. ISBN 80-7254-837-9
4. Ehara, Y., Fujimoto, H., Miyazaki, S., Mochimaru, M., Tanaka, S., Yamamoto, S.: Comparison of the performance of 3D camera systems II. Gait Posture **5**, 251–255 (1997)
5. Richards, J.G.: The measurement of human motion: a comparison of commercially available systems. Hum. Mov. Sci. **18**, 589–602 (1999)
6. Liu, H., Holt, C., Evans, S.: Accuracy and repeatability of an optical motion analysis system for measuring small deformations of biological tissues. J. Biomech. **40**, 210–214 (2007)
7. Cohen, L., Manion, L., Morrison, K.: Research Methods in Education. Routledge, London (2013)
8. Kenyon, C., Cala, S., Yan, S., Aliverti, A., Scano, G., Duranti, R., et al.: Rib cage mechanics during quiet breathing and exercise in humans. J. Appl. Physiol. **83**, 1242–1255 (1997)
9. De Groote, A., Wantier, M., Chéron, G., Estenne, M., Paiva, M.: Chest wall motion during tidal breathing. J. Appl. Physiol. **83**, 1531–1537 (1997)
10. Kaneko, H., Horie, J.: Breathing movements of the chest and abdominal wall in healthy subjects. Respir. Care **57**, 1442–1451 (2012)
11. Ragnarsdóttir, M., Kristinsdóttir, E.K.: Breathing movements and breathing patterns among healthy men and women 20–69 years of age. Respiration **73**, 48–54 (2006)
12. Kristjánsdóttir, Á., Ragnarsdóttir, M., Hannesson, P., Beck, H.J., Torfason, B.: Respiratory movements are altered three months and one year following cardiac surgery. Scand. Cardiovasc. J. **38**, 98–103 (2004)
13. Rostkowska, E., Bak, M., Samborski, W.: Body posture in women after mastectomy and its changes as a result of rehabilitation. Adv. Med. Sci. **51**, 287–297 (2006)

14. Crosbie, J., Kilbreath, S.L., Dylke, E., Refshauge, K.M., Nicholson, L.L., Beith, J.M., et al.: Effects of mastectomy on shoulder and spinal kinematics during bilateral upper-limb movement. Phys. Ther. **90**, 679–692 (2010)
15. Arndt, V., Stegmaier, C., Ziegler, H., Brenner, H.: A population-based study of the impact of specific symptoms on quality of life in women with breast cancer 1 year after diagnosis. Cancer **107**, 2496–2503 (2006)
16. Hayes, S.C., Rye, S., Battistutta, D., Newman, B.: Prevalence of upper-body symptoms following breast cancer and its relationship with upper-body function and lymphoedema. Lymphology **43**, 178–187 (2010)
17. McLaughlin, S.A., Wright, M.J., Morris, K.T., Giron, G.L., Sampson, M.R., Brockway, J.P., et al.: Prevalence of lymphedema in women with breast cancer 5 years after sentinel lymph node biopsy or axillary dissection: objective measurements. J. Clin. Oncol. **26**, 5213–5219 (2008)
18. Lewit, K., Olsanska, S.: Clinical importance of active scars: abnormal scars as a cause of myofascial pain. J. Manip. Physiol. Ther. **27**, 399–402 (2004)

Impedimetric Method to Monitor Biological Layer Formation on Central Venous Catheters for Hemodialysis Made of Carbothane

Ewa Paradowska[1]($\boxtimes$), Marta Nycz[1], Katarzyna Arkusz[1], Bartosz Kudliński[2], and Elżbieta Krasicka-Cydzik[1]

[1] Biomedical Engineering Division, University of Zielona Gora, Licealna 9 Street, 65-547 Zielona Gora, Poland
e.paradowska@ibem.uz.zgora.pl

[2] Department of Intensive Care, Zielona Gora University Hospital, Zyty 6 Street, 65-046 Zielona Gora, Poland
https://www.zib.wm.uz.zgora.pl

Abstract. The aim of the study was to specify by an impedimetric method the changes observed on the inner wall of central venous catheters for hemodialysis leading to the formation of a biological film. To evaluate these changes a patient-dialyzer model was built in which experimental parameters were kept closely similar to the clinical conditions of hemodialysis. The impedance spectra and SEM/EDS analysis of the biological layer deposited on the inner surface of the distal part of the catheter gave an insight into the structure of film formation and its chemical composition. Since an early detection of biofilm formation inside the distal part of the catheter is crucial for the safety of medical treatment and it usually prompts the implementation of antibiotic therapy. Developed impedimetric method can minimize the risk of infection and ensure the continuity of treatment.

Keywords: Central venous catheters
Electrochemical impedance spectroscopy · Biological layer
Scanning electron microscopy

1 Introduction

Recently, the number of patients undergoing hemodialysis treatment with central venous catheters [1–3] has grown. For instance, in the US, it amounts to about 6–150 million each year [4–6]. Insertion of the catheter is an invasive procedure associated with infectious and thrombotic complications [1,3,7–13], predisposing to the occurrence of a number of complications. It is estimated that every year 12–25% of patients die due to complications and infections [14,15]. Moreover, the number of such cases increases approximately by 1.5% per year

© Springer Nature Switzerland AG 2019
K. Arkusz et al. (Eds.): BIOMECHANICS 2018, AISC 831, pp. 45–55, 2019.
https://doi.org/10.1007/978-3-319-97286-2_4

[15] and the cost of complications amounts to more than 2.3 billion dollars [16]. The majority of pathogens that cause infections are microorganisms inhabiting the skin, mainly *Staphylococcus epidermidis, Staphylococcus aureus, Escherichia coli, Pseudomonas aeruginosa, Enterococcus spp., Acinetobacter spp., Klebsiella pneumonae, Enterobacter cloaca* [2,10]. Thromboses are formed by molding on the surface of the catheter a fibrin sheath comprised of numerous proteins, including albumin, fibrin, fibronectin, collagen, and laminin [12]. As a result of the formation of a bacterial film serious human catheter-related bloodstream infections (CRBSI) are observed. Thrombosis can cause problems with blood withdrawal and contribute to the partial or full occlusion [8]. A too-late diagnosis of catheter blockage is most risky during the treatment [7]. Clinical symptoms are often nonspecific or characteristic for other diseases [7]. Therefore, for a long time they are not linked to the catheter infection. It leads to increased costs, interruption of therapy as well as it adversely affects the patient's condition [4,8]. Current prevention methods which include modifications of the inner surface by antithrombotics and antibacterial agents [2,7,15–18] are so far not sufficiently effective [1]. Only recently [19] the use of the electrochemical impedance spectroscopy has led to the development of a biosensor which can monitor the formation of a biofilm layer at the port of a venous catheter, located just under the skin. The catheter with a port similar to a tunneled catheter left entirely under the skin helps to prevent the infection which is often observed in catheters described by the authors of this study. However, such type of catheter has a different construction (the type and shape of the side holes and the tip). According to numerous scientific reports [20,21] and the present study the most threatened part is the tip of the catheter, which is placed in the right atrium of the heart and thus is in constant contact with blood. The fibrin sheath formed mainly by proteins on the surface of the catheter within 24 h of its insertion into the patient's circulation promotes the subsequent adhesion of pathogens [1]. However, numerous scientists [19,22–25] used bacterial cultures (neglecting the influence of the proteins) for providing a characterization of biofilm growth, e.g. Paredes et al. used *S. epidermidis* [19,22] and *S. aureus* [23], Ben-Yoav et al. used *E. coli* [24], and Taeyoung et al. used *Pseudomonas aeruginosa* [25]. Examinations of the deposition and the growth of biofilm were recorded for frequencies range from 0.01 Hz to 400 kHz with an AC amplitude of 10 to 100 mV [19,22–25]. In most of the studies a standard three-electrode configuration [24,25] or a label-free interdigitated electrode (IDAM) biosensor were used [19,22,23]. The aim of this research was to develop an impedimetric method to monitor the initial conditioning layer formed on the inner surface of a hemodialysis catheter made of carbothane. The process of a biological layer formation was monitored for the tip of the catheter which is most vulnerable to the formation of biofilm structures. The bovine serum albumin was used as a biological factor. This protein is a part of the fibrin sheat which forms on the surface of the catheter within 24 h from placing it into the bloodstream. In order to monitor film formation by the electrochemical impedance spectroscopy (EIS) method and to check its performance, a patient-dialyser system was elaborated. The film layer formation was analyzed during the flow of the PBS solution (0.01 M, pH 7.4) containing

albumins, which account for about 60% of the total protein in human blood. The bovine serum albumin (BSA) has a negative net charge (isoelectric point - $I_{ep} = 4.7$) in human serum, similarly to the bacteria causing catheter-related bloodstream infections. Due to the above and a positive charge of the tested surface the bovine serum albumin was used as a biological factor.

2 Materials and Methods

The central venous catheters type MAHURKAR Maxid^{TM}, Covidien company, were used. This type of catheter is built with a double lumen catheter and laser cut side holes at tip. The process of biofilm formation was monitored on the distal tip of the catheter. It is made of polycarbonate-based thermoplastic polyurethanes - carbothane. The phosphate buffered saline (PBS, 0.01 M, pH 7.4), bovine serum albumin (BSA), glutaraldehyde and acetone were purchased from Sigma-Aldrich. In order to simulate clinical conditions, the system parameters were kept closely similar to clinical conditions of hemodialysis, i.e. pH = 7.4, temperature of solution - $36.6 \pm 0.1°C$ and flow rate of solution through the dialysis apparatus (Fresenius Medical Care 4008 s) - 300 ml/min.

2.1 Electrochemical Measurement

The open circuit potential (OCP) values and electrochemical impedance spectroscopy (EIS) scans were recorded using a standard three-electrode configuration, with a platinum wire as a working electrode, a standard Ag/AgCl silver chloride electrode ($E_{Ag/AgCl} = 0.222$ V) as a reference electrode, and a platinum electrode as an auxiliary electrode. The electrodes were placed in the arterial lumen of the catheter. The Atlas 0531 potentiostat/galvanostat was used for all tests.

All electrochemical measurements were performed in the PBS solution (0.01 M, pH 7.4) containing 200 mg/ml BSA for different times of flow ranging from 0 h ÷ 5 days. The measurement of the OPC values were recorded for 300 s. The EIS spectra were recorded for frequencies from 100 kHz to 0.1 Hz at zero DC potential with an AC amplitude of 10 mV.

2.2 Microscopic Analysis of Catheters

The microscopic analysis SEM/EDS of the brand new catheter and the catheter tested by the EIS rendered it is possible to compare changes on surfaces of catheters resulting from the flow of the PBS/albumin solution. Additional energy dispersive X-ray spectroscopy (EDS) analyses provided information on the elemental composition of every sample.

The microscopic analysis was performed by the scanning electron field emission microscope JEOL JSM 7600F equipped with an X-ray analyser INCA OXFORD. The microscopic observation of the biological layer required the use of an additional sample preparation procedure which includes: the immersing of samples in a 25% solution of glutaraldehyde in a phosphate buffer (pH=7.2) and rinsing (3 times) in a phosphate buffer solution at room temperature. Samples were then dehydrated in 10 ml portions of acetone-water solutions with concentration rising from 10% to 100%. Drying was carried out at the critical point of CO_2, using the critical point E3000/E3100 drying apparatus (CPD). The catheter surface was covered with a chromium layer with the thickness of 2 nm.

The roughness measurements of the brand new catheter and catheter immersed in the PBS/BSA solution for 5 days were performed by an atomic force microscope (AFM) Nanosurf EasyScan 2. The catheters were subjected to the AFM scanning in air using static mode with setpoint 20 nN. Optimal scanning parameters determined experimentally to avoid protein damage or tip contamination were as follows: scan rate, 0.4 line/s with 256 points per line; integral gain. The measurements were carried out in three selected locations on the tip of the catheter.

3   Results and Discussion

3.1   The Open Circuit Potential Values (OCP)

The stability of the electrode potential is evaluated by recording the OCP in the PBS solution (0.01 M, pH = 7.4) before and after addition of the BSA (200 mg/ml) in the standard three-electrode configuration described in the Materials and methods section for 300 s.

Figure 1 presents the OCP curves recorded for different times of flow, i.e., 0 h ÷ 5 days. The open circuit potential of the central venous catheters made

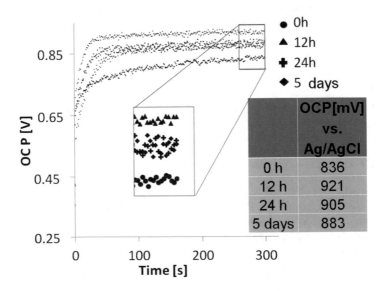

Fig. 1. Open circuit potential values (OCP) measured for 300 s in the PBS/albumin solution (0.01 M, pH 7.4) for different times of flow.

of carbothane covered with barium sulphate was 839 mV, and its values are compatible with the measurement carried out by Wei [26].

During the first 12 h of immersion of the catheter in the PBS/BSA solution, the shift of the OCP to more positive values, from 836 mV to 921 mV, was observed. However, during the period from 1 day to 5 days a decrease in the measured values, from 905 mV to 803 mV, was noticed. The OCP values increase as a result of the additional layer formation on the catheter surface, which consists of the components of the PBS solution (mainly K^+ and Na^+).

On the other hand, a slight decrease in the OPC values is related to the phosphate (PO_4^{3-}) adsorption, which is favourable in the BSA adsorption mechanism [27]. The presence of the above-mentioned elements was confirmed in further EDS analyses. A decrease in the OPC values recorded on days 1–5 confirms the adsorption of the bovine serum albumin on the catheter surface. The BSA has an I_{ep} of 4.7–5.2 and is hence negatively charged in a pH neutral fluid such as the PBS [28]. I_{ep} indicates the physisorption and causes a decrease in positively charged potential of the surface. In addition, small current oscillations seen on the recorded OPC curves may suggest ongoing deposition processes and prove the metastable nature of biological layers.

In general, the OCP values reflect not only the electric properties of the electrode but also the potential change due to the oxidation or reduction of immobilized organic molecules on the electrode surface. On the basis of the above considerations, it can be stated that the first stage of biofilm formation consists of two steps: first, the ions adsorption and then biomolecules adsorption.

3.2   Impedance Characteristics

The EIS spectra were recorded for frequencies from 100 kHz to 0.1 Hz at zero DC potential with an AC amplitude of 10 mV in the PBS/BSA solution (0.01 M, pH 7.4). Figure 2 presents the Nyquist plots for film layers and their course correlates with open circuit potential measurements.

Because of the Pt/Pt configuration in contact with the polyurethane films, it shows only one semi-circle in the Nyquist plots throughout the whole immersion process. In the initial 12 h, the resistivity of the platinum electrode increases due to the sedimentation of K^+ and Na^+ ions on the surface of catheters. With the immersion testing going on from day 1 to 5, the impedance and the resistance of the layers increase with the increase of flow time. Analyzing the results in Fig. 4 one can see that the impedance increased over time because of the BSA fouling on the electrode surface, which induces an increase in the charge transfer resistance. The above relation is well known in the literature [29,30].

Bode spectra (Fig. 3), illustrating the relation between phase angles and frequencies, show that for the lowest frequency 0.1 Hz the phase angles range from 18 to 35°. The highest heterogeneity of the tested surface (number of components of the PBS solution) was observed for the flow lasting 12 h (phase angle 35°). A decrease in the phase angle values from 28 to 19° for the flow lasting from 1 to 5 days is related to the formation of a protein layer. Due to the size of albumin (3.8 nm in diameter, 15 nm long) the surface heterogeneity decreases [31,32].

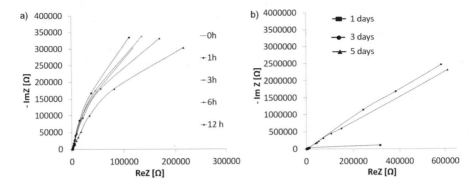

Fig. 2. Nyquist plots recorded in the PBS solution (0.01 M, pH 7.4), the frequency $0.1 - 10^5$ Hz with amplitude 10 mV vs. $E_{Ag/AgCl} = 0.222$ V, for different flow times a) spectra for initial measurement and for flows lasting from 1 to 12 h, b) spectra recorded for days 1–5 of flow through the catheters.

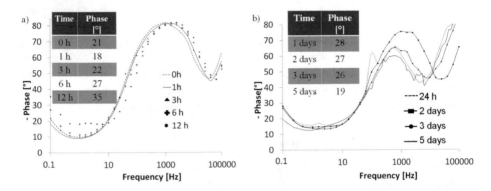

Fig. 3. Bode plots recorded in the PBS solution (0.01 M, pH 7.4) over the frequency range $0.1 - 10^5$ Hz with amplitude 10 mV vs. $E_{Ag/AgCl} = 0.222V$, for different flow times a) initial measurement and 1–12 h b) 1–5 days.

3.3   Microscopic Analysis of the Surface of Catheters

After the electrochemical experiment the analyzed samples were inspected microscopically in order to confirm the presence of BSA and PBS components. Figure 4 shows the microstructures of surface layers deposited inside the catheters. SEM images illustrate the deposition of fine precipitates covering the inner surface of the vascular catheter. The layer spreads over the venous and arterious lumen and the central partition of the catheter. The EDS analysis (Table 1), confirms that the layer consists of deposits of the solution of phosphate buffered saline (PBS) components, i.e.: carbon (C), oxygen (O), sodium (Na), potassium (K), and phosphorus (P).

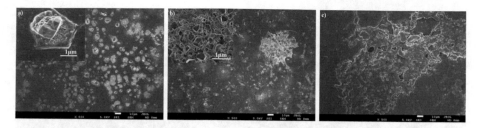

Fig. 4. SEM images of the layer on the inner surface of the catheter for its different parts characterised by the impedance spectroscopy method: a) venous lumen, b) arterial lumen, c) the central partition.

Table 1. EDS analysis of the structure of the catheter characterized by the electrochemical impedance spectroscopy.

Part of the catheter	SEM	Element	Weight [%]
Venous lumen		O Na P K	42.91 37.45 5.98 13.66
Arterial lumen		C O Na P	42.23 8.55 16.10 33.12
Central lumen		C O P	28.37 53.31 18.32

The micrographs of the brand new vascular catheter (Fig. 5) showed that the tested surface had many irregularities. The EDS analysis presented in Table 2 confirms that it is covered with barium sulfate ($BaSO_4$) - the compound used as a contrast agent which enables to localize the catheter tip after placing it in the patient's bloodstream.

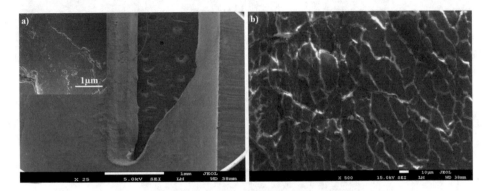

Fig. 5. SEM images of the microstructure of the brand new catheter: a) x25, b) x500.

Table 2. EDS analysis of the structure of the brand new catheter.

Part of the catheter	Element	Weight [%]
	O	28.18
	C	45.76
Spectrum 2	S	4.52
	Cr	3.07
	Ba	18.47

The AFM analysis showed that the catheter studied by EIS had the highest average roughness ($R_a = 19.61 \pm 1.96$ nm) compared to the brand new catheter ($R_a = 17.99 \pm 5.55$ nm). Higher values of R_a confirm electrochemical characteristics and analysis performed on the SEM images.

4   Conclusions

The results of the impedance analysis combined with microscopic observation confirmed the possibility of monitoring the growth of the initial biological layer on the inner surface of the catheter tip in vivo. The layer consists of carbon, oxygen, sodium, potassium and phosphorus, i.e. the components of the solution flowing through the catheter.

The surface of the brand new catheter is covered with a heterogeneous layer of barium sulfate. Such an irregular coating favours the adhesion of components of the solution used during hemodialysis as well as pathogens and biological elements. The most important is that the elaborated method offers the chance of early detection of biofilm formation in vivo, thus giving the opportunity to protect against the infection during long-term use of catheters.

References

1. Chan, M.R.: Hemodialysis central venous catheter dysfunction. Semin. Dial. **21**, 516–521 (2008). https://doi.org/10.1111/j.1525-139X.2008.00495.x
2. Saxena, A.K., Panhotra, B.R.: Haemodialysis catheter-related bloodstream infections: current treatment options and strategies for prevention. Swiss Med. Wkly. 127–138 (2005)
3. Venturini, E., Becuzzi, L., Magni, L.: Catheter-induced thrpmbosis of the superior vena cava. Vasc. Med. 1–4 (2012). https://doi.org/10.1155/2012/469619
4. O'Grady, N.P., Alexander, M., Burns, L.A., Dellinger, E.P., Garland, J., Heard, S.O.: Guidelines for the prevention of intravascular catheter-related infections. Clin. Infect. Dis. **52**, 1087–1099 (2011). https://doi.org/10.1093/cid/cir257
5. Tsukashita, M., Anda, A., Balsam, L.: Type A aortic dissection: a rare complication of central venous catheter placement. J. Card. Surg. **29**, 368–370 (2014)
6. Deliberato, R.O., Marra, A.R., Correa, T.D., Martino, M.D.V., Correa, L., Pavao dos Santos, O.F., Edmond, M.B.: Catheter related bloodstream infection (CR-BSI) in ICU Patients: making the decision to remove or not to remove the central venous catheter. PLoS OnE **7**, 3 (2012). https://doi.org/10.1371/journal.pone.0032687
7. Raad, I., Hanna, H., Maki, D.: Intravascular catheter-related infections: advances in diagnosis, prevention, and management. Lan. Infect. Dis. **7**, 645–657 (2007)
8. Vascular Access Work Group. Clinical practice guidelines for vascular access. Am. J. Kidney Dis. 176–273 (2006). https://doi.org/10.1053/j.ajkd.2006.04.029
9. Menglin, T., Mei, F., Lijun, C., Jinmei, Z., Peng, K., Shuhua, L.: Closed blood conservation device for reducing catheter-related infections in children after cardiac surgery. Crit. Care Nurse **5**, 53–61 (2014)
10. Horvath, R., Collignon, P.: Controlling intravascular catheter infections. Aust. Prescr. **26**, 41–43 (2003)
11. Elliott, T.S.J.: The pathogensis and prevention of intervascular catheter infections Central Venous Catheters. Lancet Infect. Dis. 206–215 (2009)
12. Cornelis van Rooden, J., Schippers, E.F., Rosendaal, F.R., Meinders, A.E., Huisman, M.V.: Infectious complications of central venous catheters increase the risk of catheter-related thrombosis in hematology patients: a prospective study. J. Clin. Oncol. **23**, 2655–2660 (2005). https://doi.org/10.1200/JCO.2005.05.002
13. Davenport, A., Ahmand, J.: Medical management of hepatorenal syndrome. Nephrol. Dial. Transplant. **1**, 34–41 (2012). https://doi.org/10.1093/ndt/gfr736
14. Gahlot, R., Nigam, C., Kumar, V., Yadav, G., Anupurba, S.: Symposium: current concepts in critical care. Int. J. Crit. Illn. Inj. Sci. **4**, 162–167 (2014). https://doi.org/10.4103/2229-5151.134184
15. Maczynska, B., Przondo-Mordarska, A.: Bloodstream infections related to venous access. Zakazenia, p. 4 (2011)

16. Mozaffari, K., Bakhshandeh, H., Khalaj, H., Soudi H.: Incidence of catheter-related infections in hospitalized cardiovascular patients. Res. Cardiovasc. Med. **2**, 99–103 (2013). https://doi.org/10.5812/cardiovascmed.9388

17. Kutner, D.J.: Thrombotic complications of centra venous catheters in cancer patients. Oncologist **9**, 207–216 (2004)

18. Krishnasami, Z., Ton, D.C., Bimbo, L., Taylor, M.E., Balkovetz, D.F., Barker, J., Allon, M.: Management of hemodialysis catheter-related bacteremia with an adjunctive antibiotic lock solution. Kidney Int. **61**, 1136–1142 (2002). https://doi.org/10.1046/j.1523-1755.2002.00201.x

19. Paredes, J., Alonso-Arce, M., Sedano, B., Legarda, J., Arizti, F., Gómez, E., Aguinaga, A., Del Pozo, J.L., Arana, S.: Smart central venous port for early detection of bacterial biofilm related infections. Biomed. Microdevices **16**, 365–374 (2014). https://doi.org/10.1007/s10544-014-9839-3

20. Kingdon, E.J., et al.: Atrial thrombus and central venous dialysis catheters. Am. J. Kidney Dis. **38**, 631–639 (2001)

21. Dua, R., James, K.N., Trerotola, S.: Significance of echocardiographically detected tip thrombus associated with central venous catheters. J. Vasc. Interv. Radiol. 313 (2013). https://doi.org/10.1016/j.jvir.2013.01.338

22. Paredes, J., Becerro, S., Arizti, F., Aguinaga, A., Del Pozo, J.L., Arana, S.: Real time monitoring of the impedance characteristics of Staphylococcal bacterial biofilm cultures with a modified CDC reactor system. Biosens. Bioelectron. **38**, 226–232 (2012)

23. Paredes, J., Becerro, S., Arizti, F., Aguinaga, A., Del Pozo, J.L., Arana, S.: Interdigitated microelectrode biosensor for bacterial biofilm growth monitoring by impedance spectroscopy technique in 96-well microtiter plates. Sensor Actuator **178**, 663–670 (2013). https://doi.org/10.1016/j.snb.2013.01.027

24. Ben-Yoav, H., Freeman, A., Sternheim, M., Shacham-Diamand, Y.: An electrochemical impedance model for integrated bacterial biofilms. Electrochim. Acta **56**, 7780–7786 (2011). https://doi.org/10.1016/j.electacta.2010.12.025

25. Taeyoung, K., Junil, K., Lee, J.H., Jeyong, Y.: Influence of attached bacteria and biofilm on double-layer capacitance during biofilm monitoring by electrochemicalimpedance spectroscopy. Wat. Res. **45**, 4615–4622 (2011). https://doi.org/10.1016/j.watres.2011.06.010

26. Wei, H., Ding, D., Wei, S., Guo, Z.: Anticorrosive conductive polyurethane multiwalled carbon nanotube nanocomposites. J. Mater. Chem. **45**, 10805–10813 (2013)

27. Ookubo, A., Nishida, M., Ooi, K., Ishida, K., Hashimura, Y., Ikawa, A., Yoshimura, Y., Kawada, J.: Mechanism of phosphate adsorption to a three-dimensional structure of boehmite in the presence of bovine serum albumin. J. Pharm. Sci. **82**, 744–749 (1993)

28. Wang, X., Herting, G., Wallindera, I., Blomberg, E.: Adsorption of bovine serum albumin on silver surfaces enhances the release of silver at pH neutral conditions. Phys. Chem. **17**, 18524–18534 (2015). https://doi.org/10.1039/c5cp02306h

29. Moradi, M., Yeganeh, H., Pazokifard, S.: Synthesis and assessment of novel anticorrosive polyurethane coatings containing an amine-functionalized nanoclay additive prepared by the cathodic electrophoretic deposition method. RSC Adv. **6**, 28089–28102 (2016). https://doi.org/10.1039/C5RA26609B

30. Mohsen, Q., SFadl-allah, S.A., El-Shenawy, N.S.: Electrochemical impedance spectroscopy study of the adsorption behavior of bovine serum albumin at biomimetic calcium - phosphate coating. Int. J. Electrochem. Sc. **7**, 4510–4527 (2012)

31. Vlasova, I., Saletsky, A.: Study of the denaturation of human serum albumin by sodium dodecyl sulfate using the intrinsic fluorescence of albumin. J. App. Spectrosc. **76**, 536–541 (2009)
32. Hanamura, K., Tojo, A., Kinugasa, S.: The resistive index is a marker of renal function, pathology, prognosis, and responsiveness to steroid therapy in chronic kidney disease patients. Int. J. Nephrol. 1-9 (2012). https://doi.org/10.1155/2012/139565

Towards Understanding of Mechanics of Hernia Managed by Synthetic Mesh in Laparoscopic Operation: A Single Case Study

Agnieszka Tomaszewska[1]([⊠]), Izabela Lubowiecka[1], and Czesław Szymczak[2]

[1] Faculty of Civil and Environmental Engineering, Gdansk University of Technology,
Narutowicza 11/12, 80-233 Gdansk, Poland
atomas@pg.edu.pl, lubow@pg.edu.pl
[2] Faculty of Ocean Engineering and Ship Technology,
Gdansk University of Technology, Narutowicza 11/12, 80-233 Gdansk, Poland
szymcze@pg.edu.pl

Abstract. In this paper a research towards understanding of mechanics of ventral hernia operated with the use of Physiomesh™ implant and SecureStrap® staples is described. Experimental and numerical studies are conducted for that purpose. Experimental works cover uni-axial tension tests of the implant samples and of the implant-staples-tissue system. Also experiments on implant-staples-tissue models, representing operated hernia, subjected to impulse pressure loading are performed. Based on that, constitutive model of the mesh has been identified and failure load of the staples has been determined. In the experiments on the operated hernia systems subjected to pressure loading safe loading level has been determined and failure modes connected to higher pressure values have been observed. Finally, in the numerical simulations of the operated hernia model, built according to FEM rules, it has been proved that failures observed experimentally result from exceeding of the load bearing capacity of the staples considered in this study.

Keywords: Mechanics in medicine · Mechanical properties
Hernia management

1 Introduction

Ventral hernia (VH) occurs when continuity of abdominal wall fascia is interrupted. In such case intestines are displaced outside abdominal cavity by internal pressure. The sickness is painful and is a life-danger. Hernia can occur as a primary defect of abdominal wall or as incisional defect. The statistics shows that 28% of patients who formerly had an operation in a region of abdominal wall suffer from subsequent hernia in the incision [1]. Recurrence rate of VH is high. It can reach even 43% of patients [2]. These factors make VH management

© Springer Nature Switzerland AG 2019
K. Arkusz et al. (Eds.): BIOMECHANICS 2018, AISC 831, pp. 56–67, 2019.
https://doi.org/10.1007/978-3-319-97286-2_5

relatively frequent operation. There are approximately 400 thousands of such operations annually in Europe and 300 thousands in the US [3]. Authors of the paper [4] claim that in the year 2006 alone \$3.2 billion was spent on ventral hernia repairs in the US. They also calculated that each 1% reduction of these operations would save US \$32 million dollars.

VH can be operated with the use of an implant or by closing the hernia orifice with the use of sutures only. The second solution is suitable for smaller hernia dimensions. Each solution should guarantee effective VH repair but there are problems in both solutions. This paper concerns the case of mesh implantation in the abdominal wall.

The role of the mesh is to reconstruct the abdominal wall. The key issue is to design implant with mechanical properties suitable for that task. There are numbers of implants available in the medical market. There are so called lightweight or heavy meshes, with large or small pores, with knitted or smooth structure. Discussion on characteristics of different meshes is presented in [5].

The authors of the present paper have made experiments and numerical modelling of VH operated with the use of Physiomesh™ (Ethicon, Somerville, NJ, USA) and dedicated stapler ETHICON SECURESTRAP® Absorbable Strap Fixation Device (Ethicon, Somerville, NJ, USA). Motivation to the study is a high recurrence rate (20%) of VH operated with the use of this system in comparison to recurrence rate of hernia managed by VentrallightTM ST Mesh (Davol Inc, Subsidiary of C. R. Bard, Inc., Warwick, RI, USA) and dedicated stapler SorbaFIX, which is reported in [6]. Also high post-operative pain is reported for Physiomesh™ in the same paper. The question arises: what are the mechanical properties of Physiomesh™ - Securestrap system that cause the observed failures.

2 Materials and Methods

2.1 Materials Subjected to the Study

Physiomesh™ is composed of a monofilament polypropylene (PP) knitted mesh coated with a monocryl (polyglecaprone 25) absorbable barrier layers. There is one layer to each side of the polypropylene mesh. A polydioxanone film binds the monocryl to the PP mesh. The mesh is presented in Fig. 1. The knitting pattern is marked by synthetic violet strip with width of 3 and 6 mm. The area of the mesh, which is covered by this strip is more stiff than other area - that can be easily sensed manually. This implant is not currently offered by the producer at the official www site.

ETHICON SECURESTRAP® Absorbable Strap Fixation Device is declared by the producer to be essentially absorbed in 12–18 months. It is 5 mm long. It is presented in Fig. 1 as well.

2.2 Constitutive Model of the Mesh

Based on the former experience of the authors in abdominal implants modelling, dense net material model has been selected as a constitutive model of the mesh

58   A. Tomaszewska et al.

(see e.g., [7]). Due to the discrete microstructure of the surgical mesh considered
in the study, the dense net model proposed in [8] and developed further as shown
in [9] was applied here. The main idea of this model is to define the substructure
which describes the behaviour of two fibre families, the warp and the weft. It is
assumed that the threads of the fabric work in the uniaxial tensile state; the force
in every thread only depends on the strain in it (weft and warp respectively); the
influence of the coating is neglected and the angle between threads can change
in the deformation process. According to these assumptions, the anisotropic
fabric can be represented by separated threads. The nature of separate thread
is considered as isotropic. The application of the dense net formulation in the
finite element modelling of fabrics includes the direction changes of the threads
during deformation.

In this model, assuming that the local coordinate system (x_i) of the finite
element has an axis parallel to one family of threads, the strains in two threads
families are linked to strains calculated for the plane stress state by the relation
(1) (see e.g., [9]).

$$\varepsilon_\xi = \begin{bmatrix} \varepsilon_{\xi_1} \\ \varepsilon_{\xi_2} \end{bmatrix} = \begin{bmatrix} 1 & 0 & 0 \\ \cos^2\alpha & \sin^2\alpha & \sin\alpha\cos\alpha \end{bmatrix} \begin{bmatrix} \varepsilon_{x_1} \\ \varepsilon_{x_2} \\ \gamma_{x_1 x_2} \end{bmatrix} = \boldsymbol{T}_{x\xi}\boldsymbol{\varepsilon}_x \tag{1}$$

where ε_ξ represents the strains in the direction of threads family ξ_α expressed
by the components of strains tensor $\boldsymbol{\varepsilon}_x(\varepsilon_{x_1}, \varepsilon_{x_2}, \gamma_{x_1 x_2})$ and α is the current angle
between fibres. Hence, the stress tensor can be defined as (2)

$$\sigma_\xi = \begin{bmatrix} \sigma_{\xi_1} \\ \sigma_{\xi_2} \end{bmatrix} = \begin{bmatrix} F_1(\xi_1) & 0 \\ 0 & F_2(\xi_2) \end{bmatrix} \begin{bmatrix} \varepsilon_{\xi_1} \\ \varepsilon_{\xi_2} \end{bmatrix} = \boldsymbol{F}\boldsymbol{\varepsilon}_\xi \tag{2}$$

Here, the constitutive functions $F_1(\xi_1)$ and $F_2(\xi_2)$ are uniaxial for the wrap
and the weft, as presented in [10]. In the surgical mesh considered in this study
the constitutive functions F_1 and F_2 for two selected directions '1' and '2' are
represented by piecewise constant functions determined by moduli of elasticity
E_1 and E_2 taking the values from the tensile tests results as shown in Sects. 2.3
and 3.

Dense net material model is to be applied in the majority of the mesh area,
because of its knitted structure. Only the area covered by the violet strip is
modelled as homogeneous, isotropic material because stiffness of this coating
determines mechanical properties of this part of the mesh.

2.3   Experiments

To identify mechanical properties of the materials under the research and
their combination, three kinds of experiments have been performed. They are
described below.

Uniaxial Tension Tests of the Mesh. The experiments are made to identify elastic moduli of the mesh in two selected directions. They will describe

stiffness functions in the constitutive model. Knitting pattern and violet strip parallel to it point at orthogonality of the material. Its different stiffness in the directions parallel (direction '1' of the mesh, say) and perpendicular (direction '2') to the violet strip can be manually sensed. Thus, two kinds of samples have been prepared for the tests, each of three pieces. They are cut out of the mesh in the directions '1' and '2'. The samples steer clear of the area covered by the violet strip. Strip alone has been prepared as well. Each sample has been 25 mm wide and at least 110 mm long to preserve 90 mm clamp-to-clamp distance during tests. The samples are visible in Fig. 1. Zwick Roell Z020 strength machine equipped with video-extensometer has been used. The samples have been stretched until rupture with constant strain rate equal to 0.001 1/s. Elongation of measuring base, equal approximately 25 mm in each test as well as force have been collected. Based on that stress-strain relations have been determined. Engineering measures have been selected.

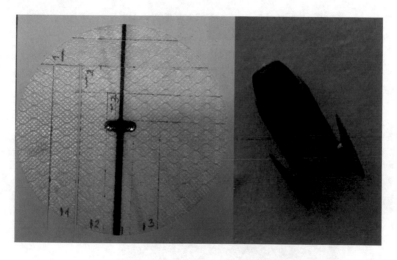

Fig. 1. (left) Physiomesh™ prepared for samples cutting; (right) ETHICON SECURESTRAP®

Uniaxial Tension Tests of the Mesh-Staples-Tissue System. Junction force in this system has been identified based on stretching tests of especially prepared samples. They are built of rectangular pieces of the mesh and porcine tissue with abdominal fascia on one side. Two SecureStraps joint the mesh and the fascia as presented in Fig. 2. The models are stretched until the junction failure. The failure force determines load bearing capacity of tissue-SecureStrap®-Physiomesh™ system.

Tension Tests of the Mesh-Staples-Tissue System Subjected to Pressure Loading. Mesh implanted to abdominal wall is subjected to multidirectional stretching due to body movements or intraabdominal pressure. To observe mechanical behaviour of PhysiomeshTM in such conditions three similar models of repaired hernia have been built and subjected to pressure loading. The models are built of a porcine abdominal wall in which orifice is cut ('hernia') with a diameter of 5 cm and of Physiomesh™, which 'repairs' the hernia. The mesh is fixed to the tissue by 12 staples. The staples have circular layout with a diameter of 12 cm, so there is 3.5 cm of tissue and mesh overlap. Each model is placed in a pressure chamber, described in [11] and dynamically loaded by an impulse of air pressure. Such loading simulates post-operational cough, which is a common cause of operated hernia system failure, when biological process of the tissue and mesh integration is just at the beginning. Displacements of the models under different pressure values in a range of 14–25 kPa are measured and failure modes are noted, if occurred. In the area of our interests are displacements of the mesh center in the direction perpendicular to the mesh plane. The example model and a scheme of the pressure chamber are presented in Fig. 3.

Fig. 2. Mesh-staples-tissue system in uniaxial tension tests

2.4   Finite Element Modelling and Simulation of the Physiomesh™ Behaviour

The mechanical behaviour of implant called Physiomesh™ has been simulated by means of Finite Element Method (FEM) using MSC.Marc® commercial system. The Physiomesh™ implant after implantation into abdominal wall has been modelled as proposed in [7]. It is a polygonal membrane supported in 12 points with appropriate boundary conditions representing the abdominal wall around the hernia orifice of 5 cm and the fascia where its joints with the abdominal wall are placed. The abdominal wall is represented here by elastic foundation and

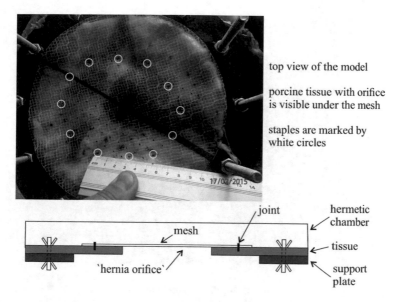

top view of the model

porcine tissue with orifice
is visible under the mesh

staples are marked by
white circles

Fig. 3. (top) Mesh-staples-tissue model for pressure loading tests; (bottom) scheme of
the pressure chamber

fascia is represented by elastic springs in the joints points defined in the plane of
the membrane (for details of the model see [7]). Four-node membrane elements
QUAD(4) with 3 translational degrees of freedom at each node are used. The
model is discretised by 848 finite elements with mesh refinement around the
tissue-implant joints. Only half of the implant is implemented using symmetry
along the centre of the violet strip (Fig. 4). Two load scenarios have been simulated, with pressure value of 14 kPa and of 25 kPa. Reaction forces have been
calculated in both cases and compared to load bearing capacity of the staples
considered in this study.

3   Results

3.1   Stiffness Functions of the Mesh

Stress-strain relations obtained experimentally are presented in Fig. 5. Multilinear approximations of the relations are shown in Fig. 6. Based on that, an
elastic modulus has been determined for each linear function. The obtained moduli determine stiffness functions of the material, necessary for dense net model
description. Two functions are determined for the majority of the mesh area, in
the '1' and '2' directions, as discussed in Sect. 2. However, the area covered by
violet strip has different stiffness, so it is modelled by a different function, the
same in two perpendicular directions '1' and '2'. The parameters of the stiffness
functions can be found in Table 1.

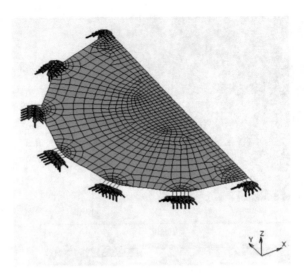

Fig. 4. Model of the implant

Table 1. Parameters of stiffness functions for '1' and '2' directions of the mesh and for violet strip.

Area, direction	Strain range	Elastic modulus value [N/mm]
Mesh, '1' direction	0–0.09	11.30
	0.09–0.60	0.79
	0.60–0.95	3.18
Mesh, '2' direction	0–0.04	18.37
	0.04–0.23	3.26
	0.23–0.45	16.57
Violet strip, '1' and '2' directions	0–0.06	30.00
	0.06–0.17	0.00
	0.17–0.39	1.31

3.2 Load Bearing Capacity of Physiomesh™ - SecureStrap® - Tissue System

The identified failure force for junction of Physiomesh™ and tissue realized by SecureStrap® equals 4 N, as identified based on experiments described in Sect. 2. Failure mode is the staple take out from the tissue.

3.3 Behaviour of Physiomesh™ - SecureStrap® - Tissue System Under Impulse of Pressure in Experiments

The first impulse pressure applied to the model has the magnitude of 25 kPa. Such pressure can occur in human abdominal cavity during jumping, as reported

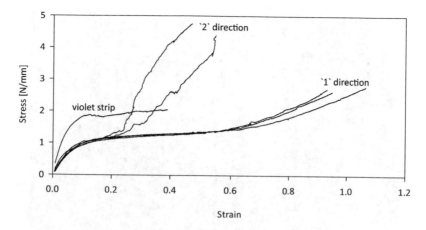

Fig. 5. Stress-strain relations obtained based on uni-axial tensile tests

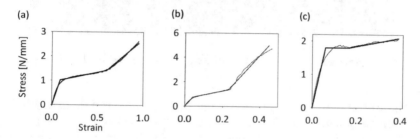

Fig. 6. Multilinear representation of stress-strain relations for (a) '1' direction of the mesh; (b) '2' direction; (c) violet strip

in [12]. The model has failed - two staples have been pulled out of the tissue and one staple has been broken. The failure mode is presented in Fig. 7. After that the model has been rebuilt. The same materials have been used by the mesh has been rotated on the meat by 90°. That has been made to check if orientation of the mesh in relation to collagen fibres orientation in the fascia matters. The importance of the mesh orientation in the abdominal wall is discussed in [13]. This model has been loaded by lower pressure value, equal to 18 kPa. Such pressure is possible during human jumping but also during coughing, according to [12]. This model has failed as well - one staple has been pulled out of the tissue and one staple has been broken.

The last model is built of the same tissue, but covered by a new piece of an implant. The pressure value is reduced again, to the value of 14 kPa. Such intraabdominal pressure is possible during formerly mentioned human activities but also during standing valsava maneuver or walking on stairs, see [12]. This load has been applied to the model four times in a row and no failure has occurred. After that an impulse pressure of 25 kPa has been realised and the model has failed in the same manner as the previously tested systems - two

Fig. 7. Failure mode of the model loaded by an impulse of 25 kPa air pressure.

staples have been pulled out of the tissue. In every case the damaged fixations are located in oblique direction in relation to the violet strip of the mesh (the approximate angle is 45°). That means, that the biggest reaction force occurs in this direction. Taking the result presented in the Sect. 3.2 one can speculate, that with the impulse pressure of 14 kPa the junction forces in the mesh fixation points do not exceed 4 N. But for the pressures with the magnitudes bigger than 18 kPa the reaction forces exceed 4 N. That supposition is verified in the subsequent section.

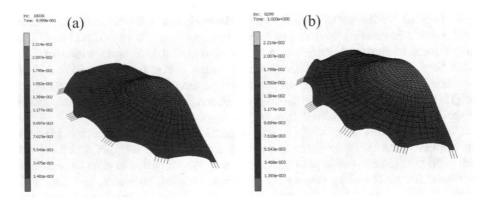

Fig. 8. Membrane displacement under 14 kPa (a) and under 25 kPa (b)

3.4   Numerical Simulations of the System Behaviour

The forces in the joints of tissue and implants, reaction forces in the FEM model, have been calculated within the numerical simulations performed under two pressure levels, 14 kPa and 25 kPa. The following maximum forces were obtained: 3.79 N and 5.29 N respectively. That means that in case of higher pressure, the force in joints exceed the capacity of the junction. The maximum displacements in both simulations were 17 mm and 23 mm respectively to the load (Fig. 8). The stress diagram (Fig. 9) also shows a clear increase of stress in the strip, where the stiffness is higher than in the rest of the membrane.

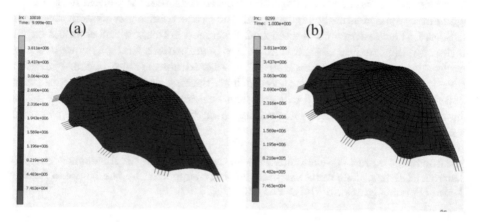

Fig. 9. Von Mises stress in model under 14 kPa (a) and under 25 kPa (b)

4   Discussion

The analysis presented in this paper concerns a complex issue of modelling of the mechanical behaviour of a system in human body based on experimental studies. The final results obtained numerically, which are the maximal reaction forces in operated hernia system, compared to load bearing capacity of the staples considered explain the mechanics of managed hernia system failures, which are observed by the doctors (see [6]). The obtained numerically results prove the correctness of FEM model of the tested system. In simulations of the experiment in which no fixation damage is observed the reaction forces do not exceed load bearing capacity of the staples, determined experimentally. On the other hand, in simulations of the experiment in which a few staples have been pulled out of the tissue the maximal reaction forces exceed the identified load bearing capacity of the considered staples.

The strengthening of the mesh with the extra strip seems to influence negatively the implanted mesh behaviour, since under pressure it causes a stress localisation along this stiffer part of the material. It may be important in particular, when it is oriented along a direction of the largest abdominal strains.

It should be also bear in mind, that mechanical properties of knitted meshes can change under cyclic loading which typically acts on the abdominal wall i.e. during coughing. The results presented in papers [14,15] prove that stiffness of the meshes increases under cyclic tension in relation to stiffness identified during simple tension. That feature should be also checked for the mesh discussed in this study, because stiffness of the mesh is a key issue influencing junction force in the mesh-tissue system, as proved in [16].

5   Conclusions

As presented in the Introduction, the considered system for ventral hernia management reveals definitely higher recurrence rate than other system discussed in Sect. 1. The research conducted and described here shows mechanical explanation for this medical observation. It has been proved, that physiological load level can cause fixation damage in the discussed here system because junction force in the fixation points can exceed load bearing capacity of the considered staples. It should be noted however, that the analysis presented here refers to the cases of newly operated hernias, in which tissue incorporation in the mesh is not started yet.

Acknowledgements. Calculations have been carried out at the Academic Computer Centre in Gdańsk. This work has been partially supported by the National Science Centre (Poland) [grant No. UMO-2017/27/B/ST8/02518].

References

1. Pans, A., Elen, P., Dewé, W., Desaive, C.: Long-term results of polyglactin mesh for the prevention of incisional hernias in obese patients. World J. Surg. **22**(5), 479–483 (1998). https://doi.org/10.1007/s002689900420
2. Burger, J.W., Luijendijk, R.W., Hop, W.C., Halm, J.A., Verdaasdonk, E.G., Jeekel, J.: Long-term follow-up of a randomized controlled trial of suture versus mesh repair of incisional hernia. Trans. Meet. Am. Surg. Assoc. **CXXII**, 176–183 (2004). https://doi.org/10.1097/01.sla.0000141193.08524.e7
3. Sauerland, S., Walgenbach, M., Habermalz, B., Seiler, C.M., Miserez, M.: Laparoscopic versus open surgical techniques for ventral or incisional hernia repair. Cochrane Database Syst. Rev. (2011). https://doi.org/10.1002/14651858.cd007781.pub2
4. Poulose, B.K., Shelton, J., Phillips, S., Moore, D., Nealon, W., Penson, D., Beck, W., Holzman, M.D.: Epidemiology and cost of ventral hernia repair: making the case for hernia research. Hernia **16**(2), 179–183 (2011). https://doi.org/10.1007/s10029-011-0879-9
5. Deeken, C.R., Lake, S.P.: Mechanical properties of the abdominal wall and biomaterials utilized for hernia repair. J. Mech. Behav. Biomed. Mater. **74**, 411–427 (2017). https://doi.org/10.1016/j.jmbbm.2017.05.008
6. Pawlak, M., Hilgers, R.D., Bury, K., Lehmann, A., Owczuk, R., Śmietański, M.: Comparison of two different concepts of mesh and fixation technique in laparoscopic ventral hernia repair: a randomized controlled trial. Surg. Endosc. **30**(3), 1188–1197 (2015). https://doi.org/10.1007/s00464-015-4329-0

7. Lubowiecka, I.: Mathematical modelling of implant in an operated hernia for estimation of the repair persistence. Comput. Methods Biomech. Biomed. Eng. **18**(4), 438–445 (2015). https://doi.org/10.1080/10255842.2013.807506

8. Branicki, C., Kłosowski, P.: Static analysis of hanging textile membranes in nonlinear approach. Arch. Civ. Eng. **29**(3), 189–220 (1983)

9. Kłosowski, P., Komar, W., Woźnica, K.: Finite element description of nonlinear viscoelastic behaviour of technical fabric. Constr. Build. Mater. **23**(2), 1133–1140 (2009). https://doi.org/10.1016/j.conbuildmat.2008.06.002

10. Ambroziak, A., Kłosowski, P.: Review of constitutive models for technical woven fabrics in finite element analysis. AATCC Rev. **11**(3), 58–67 (2011)

11. Tomaszewska, A., Lubowiecka, I., Szymczak, C., Śmietański, M., Meronk, B., Kłosowski, P., Bury, K.: Physical and mathematical modelling of implant-fascia system in order to improve laparoscopic repair of ventral hernia. Clin. Biomech. **28**(7), 743–751 (2013). https://doi.org/10.1016/j.clinbiomech.2013.06.009

12. Cobb, W.S., Burns, J.M., Kercher, K.W., Matthews, B.D., Norton, H.J., Heniford, B.T.: Normal intraabdominal pressure in healthy adults. J. Surg. Res. **129**(2), 231–235 (2005). https://doi.org/10.1016/j.jss.2005.06.015

13. Lubowiecka, I., Szepietowska, K., Szymczak, C., Tomaszewska, A.: A preliminary study on the optimal choice of an implant and its orientation in ventral hernia repair. J. Theor. Appl. Mech. **54**, 411–421 (2016). https://doi.org/10.15632/jtam-pl.54.2.411

14. Li, X., Kruger, J.A., Jor, J.W., Wong, V., Dietz, H.P., Nash, M.P., Nielsen, P.M.: Characterizing the ex vivo mechanical properties of synthetic polypropylene surgical mesh. J. Mech. Behav. Biomed. Mater. **37**, 48–55 (2014). https://doi.org/10.1016/j.jmbbm.2014.05.005

15. Tomaszewska, A.: Mechanical behaviour of knit synthetic mesh used in hernia surgery. Acta Bioeng. Biomech. **18**(1), 77–86 (2016). https://doi.org/10.5277/ABB-00185-2014-03

16. Szymczak, C., Lubowiecka, I., Tomaszewska, A., Smietanski, M.: Modeling of the fascia-mesh system and sensitivity analysis of a junction force after a laparoscopic ventral hernia repair. J. Theor. Appl. Mech. **48**(4), 933–950 (2010). http://ptmts.org.pl/jtam/index.php/jtam/article/view/v48n4p933/283

Numerical Analysis of the Blood Flow in an Artery with Stenosis

Michał Tomaszewski$^{(\boxtimes)}$ and Jerzy Małachowski

Military University of Technology,
Gen. Witolda Urbanowicza 2, Warsaw, Poland
{michal.tomaszewski,jerzy.malachowski}@wat.edu.pl

Abstract. The study presents the results of a simulated blood flow in an artery affected with arteriosclerosis. Distribution of flow velocity vectors is presented along with distribution of shear strain on the walls of the artery. During the stent designing process, the knowledge about a pathophysiological role of the shear strains during the restenosis process and about the possible phlebitis is required. According to many studies, low shear strain levels are connected with the forming of the atherosclerotic plaques with an irregular structure. The study aims to present a process that will allow obtaining numerical models of the vessel and the atherosclerotic plaque from photographs of the cross sections made with medical equipment. Those models were later used to develop a domain of the blood flow inside the vessel. The analysis was conducted using the Finite Volume Method in Ansys Fluent software. This methods converts the differential equations into algebraic ones by integrating those equations at the limits of each finite volume. The constant development of the materials and manufacturing processes for the stents allows for improvement of their usability, however one factor is not still diagnosed adequately, namely the restenosis - a condition in which a vessel undergoes narrowing again after the treatment. The rapid progress of computer methods allows for simulating increasingly complex scenarios, which can help improve the medical treatment procedures.

Keywords: Fluent · Stenosis artery · Blood flow · Arteriosclerosis
TAWSS · Restenosis · Atherosclerosis

1 Introduction

Each part of the body requires a constant supply of oxygen and nutrients. An appropriate amount of blood is, thus, required to keep the organism in running condition. It is the blood that functions as a transporting vessel for the body. The rapid life pace along with increasing stress levels and an inappropriate diet lead to many circulatory system diseases, such as arteriosclerosis causing the blood vessels to become narrowed. Arteriosclerosis is a drawnout process of building up detrimental products of metabolism inside the vessels, such as cholesterol. The buildup can cause a complete obstruction of the vessel, blocking the blood flow and causing heart stroke. In order to restore a regular blood flow in the afflicted vessel, an angioplasty procedure [1] is performed to insert a stent into the vessel. Numerical methods allow simulating the

© Springer Nature Switzerland AG 2019
K. Arkusz et al. (Eds.): BIOMECHANICS 2018, AISC 831, pp. 68–77, 2019.
https://doi.org/10.1007/978-3-319-97286-2_6

blood flow in various different vessel types [2]. A hemodynamic phenomenon of the blood flow is one of the key phenomena to the restenosis process [3, 4]. Through numerical fluid mechanics it is possible to study the fluid behavior in a scale much smaller than an experimental approach allows for. Various methods for performing angioplasty and different stent models are studied by many scientists [5–8]. An undeniable advantage numerical methods is a possibility to simulate various different vessels, allowing for analysis of the vessel before and after performing angioplasty [6, 9]. The most optimal methods for performing angioplasty are still a problem to be solved. The behavior of the vessel after angioplasty depends on many factors, such as a stent itself [10], stent expansion process [11] (kissing balloon, one, two stents), atherosclerotic plaque morphology [12], plaque shape and composition, damage to the vessel during the angioplasty, or disruption of the flow caused by protruding parts of the stent. The researchers suspect another cause of the restenosis to be in the thickness of the neointima, which can influence the strains in the walls of the vessels after performing angioplasty and local changes to the fluid dynamics caused by the presence of the stent.

2 Geometry of the Artery

2.1 Vessel Geometry Divided by Layers

In order to build a flow domain for a vessel with realistic narrowing, a geometrical model based on the medical examination was used. To developed the geometrical model, the authors used cross sections of the exemplary coronary artery. Human arteries are composed of three layers, called intima, media and adventina. Each differs from others in structure and properties. The whole geometry was developed in Autodesk Inventor software. During the development process, three basic areas were highlighted - vessel lumen, area of atherosclerotic plaque buildup, and actual artery wall (Fig. 1).

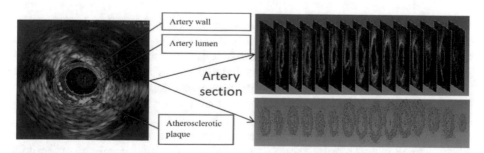

Fig. 1. Development of the artery cross sections from the images obtained from endovascular echo

The creation process of the whole structure of the atherosclerotic plaque and the vessel wall was divided into several steps. first one was placement of 15 images obtained from IVUS examination, parallel to each other and with 1.5 mm spacing between them.

The second step was to outline the vessel cross sections in order to extrude the three earlier mentioned wall layers in the third step. The next step was subtraction of the used geometries in order to obtain separated geometries of the plaque (Fig. 2) and the actual artery wall (Fig. 3).

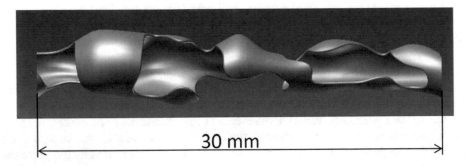

Fig. 2. Actual geometry of the atherosclerotic plaque redeveloped from the cross sections obtained from the endovascular echo examination

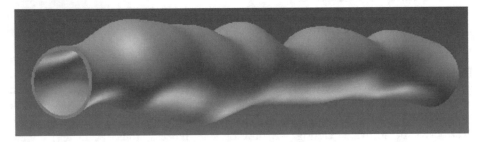

Fig. 3. Actual geometry of the artery wall redeveloped from the cross sections obtained from the endovascular echo examination

As a result of the procedure, a 30 mm fragment of the atherosclerotic plaque was obtained. It is characterized by stochastic shapes and completely asymmetrical and also has uneven thickness of the cross section. Apart from the plaque geometry, the coronary artery wall geometry was also obtained, with uneven thickness on the same 30 mm fragment. Later, the wall geometry was divided into three layers – outer and inner membranes with thickness of 0.02 mm and a medial membrane with varying thickness.

In order to develop a flow domain, the geometry needed to be modified to fill the empty volume with elements. These elements would simulate the volume of the fluid inside the artery. The first step was to simplify the geometry. Sharp transitions in the vessel were removed, as they could cause the elements to degenerate. Such elements would prevent the flow simulation from obtaining convergence. Simplification of the transition between the atherosclerotic plaque and the healthy vessel was obtained by creating a surface between the curves. The final geometry contained smooth transitions between surfaces (Fig. 4).

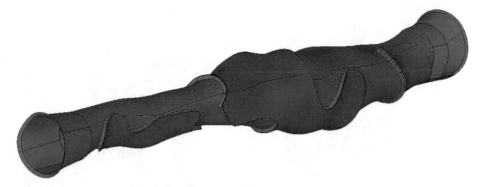

Fig. 4. Transition between the atherosclerotic tissue and the healthy vessel

3 Discrete Model

Based on the aforementioned geometrical model, a discrete model was prepared. First, the surface was meshed with tria elements, which are shell elements with three integration points. Certain parts required a thicker mesh in order to properly reflect the curvature of the vessel (Fig. 5). In order to stabilize the velocity profiles, the entry and exit parts of the vessel were extended.

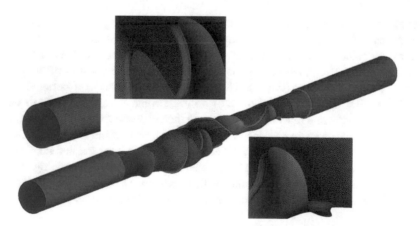

Fig. 5. 2D discrete model with actual narrowing

After checking the deformation of the elements, a 2D mesh was used to generate a 3D mesh using the CFDTetraMesh algorithm from HyperMesh software [13]. This algorithm fills the closed volume defined by tria elements with tetragonal elements. The elements were generated using the Interpolate option, causing a smooth transition from the smallest elements to the biggest ones with an initially defined size. The final 3D discrete model consisted of around 6 818 883 tetragonal finite volumes (Fig. 6).

Fig. 6. 3D discrete model with actual narrowing

3.1 Boundary Conditions

In order to perform the analysis, boundary conditions need to be specified. In the case of the blood flow in a narrowed vessel analyzed in this paper, the following boundary conditions are required:

- initial velocity,
- exit pressure,
- friction between the fluid and the walls,
- viscosity as a function of shear strain.

In the actual blood flow, the pressure is not constant. Average systolic pressure tends to be around 120 mm Hg, while the diastolic pressure averages around 80 mm Hg [14]. The pressure value as a function of time was applied to the surface outlet as a "pres-sure-outlet" condition. The blood velocity is also variable in time because of the pulsating character of the heart beat (Fig. 7). The velocity was defined as a periodic function through the UDF (user defined function) option, which allows the user to define a custom variable in a time function. In the case of the vessels with atherosclerosis, sudden changes in the cross section can occur, reducing the lumen of the vessel. In those areas the blood flow becomes turbulent and cannot be treated as a Newtonian fluid. Owning to this fact, the blood was analyzed as a non-Newtonian fluid. For defining the blood properties, a modified Power Law model proposed by [15] was used, in which the viscosity is dependent on the shear rate and the level of shear strains.

$$
\begin{cases}
\eta = 0,55471 \ Pa \cdot s \ for \ \frac{\partial V}{\partial y} < 10^{-9} s^{-1} \\
\eta = \eta_0 \left(\frac{\partial V}{\partial y}\right)^{n-1} \ for \ 10^{-9} \leq \frac{\partial V}{\partial y} < 327 s^{-1} \\
\eta = 0,00345 \ Pa \cdot s \ for \ \frac{\partial V}{\partial y} \geq 327 s^{-1}
\end{cases}
\tag{1}
$$

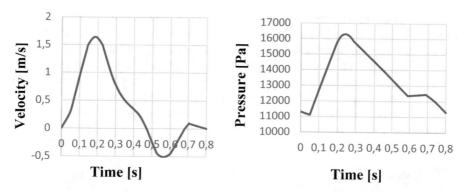

Fig. 7. Velocity and pressure profile using in analysis

4   Results

As a result of the performed analysis, the velocity and pressure distribution maps were obtained for each domain. The analysis of the blood flow in time was performed using a coupling method with the k-omega SST model of turbulence [16]. A second order integration scheme was chosen both for the momentum and the implicit analysis. The convergence criteria were set as 10e-5 for the continuum and 10e-6 for the residual velocity. The time step was set as 0,004 (It was aimed to obtain 200 steps for one full cycle of the heartbeat, which lasts 0.8 s). After obtaining the 3D visualization of the blood flow during the systolic heart phase, it becomes apparent that the streams in the center of the vessel present the highest velocity, reaching 7 m/s in areas with the most severe narrowing of the vessel (Fig. 8).

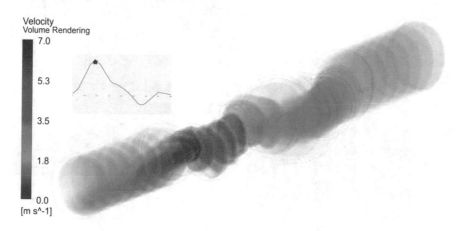

Fig. 8. Volume rendering velocity

Visualization of velocities inside the vessel were obtained by creating additional surfaces on which the values could be displayed. As it is seen in Figs. 9 and 11, local drops in pressure can be observed when the cross section area changes. Distribution of velocity vectors provides useful information (Fig. 10). It is possible to observe that the blood flowing through such a geometry there occur many areas of a turbulent flow, where substances building the arteriosclerosis tissue would accumulate over time on the vessel walls. The areas of lower velocity can be also observed after passing the narrowed parts.

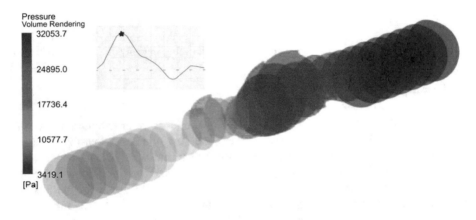

Fig. 9. Volume rendering pressure

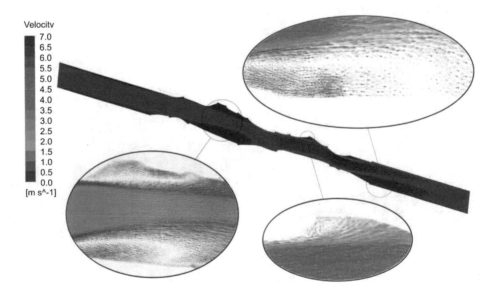

Fig. 10. Velocity map for the systolic phase

Fig. 11. Pressure map for the systolic phase

As it can be observed from the map (Fig. 12), the highest shear stress values reach 190.5 Pa. Endothelial cells affected by WSS greater than 1 Pa tend to elongate and set in the direction of the flow, while the cells affected by WSS values lower than 0.4 Pa or oscillating ones are more round and do not show any regularities. Those round cells, together with the blood stagnation often observed in the areas with low WSS, can lead to an increased walls penetration by the hematogenous particles, a phenomenon frequently associated with arterial diseases. It is a consequence of both the increased time the blood particles spend in given area and the increased permeability of the endothelium.

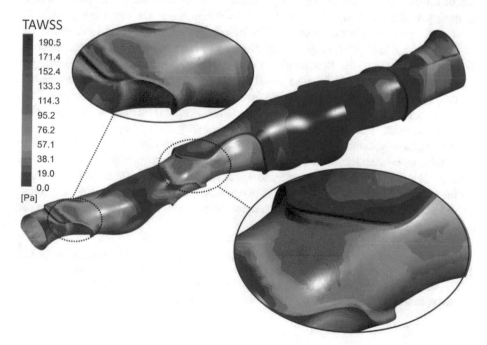

Fig. 12. Time average wall shear stress TAWSS

The WSS vector can be defined as [6]:

$$\tau_w = n \cdot \overline{\overline{\tau_{ij}}} \tag{2}$$

Where: n- vector normal to the surface, τ_{ij} - stress tensor in fluid, its length being equal to the shear stress on the surface, and its direction is the direction in which the stresses caused by the fluid viscosity affect the surface.

For the flow as a function of time, TAWSS (time-average WSS) can be defined as:

$$TAWSS = \frac{1}{T} \int_{0}^{T} |\tau_w| dt \tag{3}$$

Where: T – work cycle duration.

5   Conclusions

Abnormal levels of shear stress can lead to irreparable changes in the vessel inner surface and cause vascular diseases. From the physiological point of view, it is important to know the levels of the shear stresses and to assess what can cause them to change. The shear stress level should be constant along the whole length of the vascular tree, with no differences between individual vessels of different sizes. The stress levels can be affected by the transitions from a laminar flow to the undesirable turbulent flow, or during vessel geometry changes. The areas where the shear stress levels drop are more susceptible to vascular diseases. The blood flow character inside the vessel can influence many new circulatory system diseases to arise, among others, the stenosis. The knowledge about the flow character can provide information about critical spots where lumen narrowing can appear. Well developed algorithms and suitable methods for examination of the blood flow inside the vessel can be important tools helping heart surgeons make decisions about possible treatments. Proper modeling of the blood flow requires accurate redevelopment of vessel geometry, boundary conditions, fluid mechanics, flow parameters, and suitable domain discretization. As it is presented in the study, those requirements are possible to be achieved through combination of angiographic and IVUS examinations and validation experiments for model calibration. Further steps of the study will involve experimental studies on determining parameters for the analysis and a final model for assessing the flow through various geometries can be developed.

Acknowledgements. The study was supported by the NCBiR within project APOLLO-STRATEGMED (2/269760/1/NCBR/2015). This support is gratefully acknowledged.

This research was performed with the support of the Interdisciplinary Centre for Mathematical and Computational Modelling (ICM) University of Warsaw under grant no. GB65-19.

References

1. Morlacchi, S., Colleoni, S.G., Cárdenes, R., Chiastra, C., Diez, J.L., Larrabide, I., Migliavacca, F.: Patient-specific simulations of stenting procedures in coronary bifurcations: two clinical cases. Med. Eng. Phys. **35**, 1272–1281 (2013). https://doi.org/10.1016/j. medengphy.2013.01.007

2. Steinman, D.A., Taylor, C.A.: Flow imaging and computing: Large artery hemodynamics. Ann. Biomed. Eng. **33**, 1704–1709 (2005). https://doi.org/10.1007/s10439-005-8772-2

3. Murphy, J., Boyle, F.: Predicting neointimal hyperplasia in stented arteries using time-dependant computational fluid dynamics: a review. Comput. Biol. Med. **40**, 408–418 (2010). https://doi.org/10.1016/j.compbiomed.2010.02.005

4. Williams, A.R., Koo, B.-K., Gundert, T.J., Fitzgerald, P.J., LaDisa, J.F.: Local hemodynamic changes caused by main branch stent implantation and subsequent virtual side branch balloon angioplasty in a representative coronary bifurcation. J. Appl. Physiol. **109**, 532–540 (2010). https://doi.org/10.1152/japplphysiol.00086.2010

5. Bukala, J., Kwiatkowski, P., Malachowski, J.: Numerical analysis of crimping and inflation process of balloon-expandable coronary stent using implicit solution. Int. J. Numer. Method. Biomed. Eng. **33**, 1–11 (2017). https://doi.org/10.1002/cnm.2890

6. Chiastra, C., Morlacchi, S., Gallo, D., Morbiducci, U., Cardenes, R., Larrabide, I., Migliavacca, F.: Computational fluid dynamic simulations of image-based stented coronary bifurcation models. J. R. Soc. Interface **10**, 20130193 (2013). https://doi.org/10.1098/rsif. 2013.0193

7. Bukala, J., Kwiatkowski, P., Malachowski, J.: Numerical analysis of stent expansion process in coronary artery stenosis with the use of non-compliant balloon. Biocybern. Biomed. Eng. **36**, 145–156 (2016). https://doi.org/10.1016/j.bbe.2015.10.009

8. Bukala, J., Malachowski, J., Kwiatkowski, P.: Finite element analysis of the percutaneous coronary intervention in a coronary bifurcation. Acta Bioeng. Biomech. **16**, 23–31 (2014). https://doi.org/10.5277/ABB-00041-2014-02

9. Chiastra, C.: Numerical modeling of hemodynamics in stented coronary arteries (2013)

10. Morton, A.C., Crossman, D., Gunn, J.: The influence of physical stent parameters upon restenosis. Pathol. Biol. **52**, 196–205 (2004). https://doi.org/10.1016/j.patbio.2004.03.013

11. Timmins, L.H., Meyer, C.A., Moreno, M.R., Moore, J.E.: Effects of stent design and atherosclerotic plaque composition on arterial wall biomechanics. J. Endovasc. Ther. **15**, 643–654 (2008). https://doi.org/10.1583/08-2443.1

12. Katritsis, G.D., Siontis, G.C.M., Ioannidis, J.P.A.: Double versus single stenting for coronary bifurcation lesions a meta-analysis. Circ. Cardiovasc. Interv. **2**, 409–415 (2009). https://doi.org/10.1161/CIRCINTERVENTIONS.109.868091

13. Tomaszewski, M., Baranowski, P., Małachowski, J., Damaziak, K., Bukała, J.: Analysis of artery blood flow before and after angioplasty. In: AIP Conference Proceedings (2018)

14. Jozwik, K., Obidowski, D.: Numerical simulations of the blood flow through vertebral arteries. J. Biomech. **43**, 177–185 (2010). https://doi.org/10.1016/j.jbiomech.2009.09.026

15. Reorowicz, P., Obidowski, D., Klosinski, P., Szubert, W., Stefanczyk, L., Jozwik, K.: Numerical simulations of the blood flow in the patient-specific arterial cerebral circle region. J. Biomech. **47**, 1642–1651 (2014). https://doi.org/10.1016/j.jbiomech.2014.02.039

16. Versteeg, H., Malalasekera, W.: An Introduction to Computational Fluid Dynamics. The Finite Volume Method Itle. Longman Scientific & Technical (1995)

Computational Imaging and Simulation Technologies in Biomechanics

Influence of Elevated Temperature During Crimping on Results of Numerical Simulation of a Bioresorbable Stent Deployment Process

Jakub Bukała, Krzysztof Damaziak, Jerzy Małachowski,
and Łukasz Mazurkiewicz(✉)

Military University of Technology,
2 Gen. W. Urbanowicza Str., 00-908 Warsaw, Poland
lukasz.mazurkiewicz@wat.edu.pl

Abstract. Bioresorbable stents (BRSs) represent a promising technological development within the field of cardiovascular angioplasty because of their ability to avoid long-term side effects of conventional stents such as in-stent restenosis, late stent thrombosis and fatigue induced strut fracture. However polymer materials used for production of some of the BRSs pose new challenges raising from the fact, that mechanical properties of polymers are very different from the metallic materials used to make stents before BRSs era. These challenges manifests not only in clinical practice but mainly in the process of design of the new device. This especially applies to Finite Element based numerical simulations of the stent structure, as the first-choice tool to examine newly developed stent in early stage of design process. In the article authors investigating different scenarios of numerical simulation of stent deployment process. The goal of the exercise is to find a proper way to model influence of elevated temperature present during crimping on the behaviour of the stent.

Keywords: Bioresorbable stent · Finite element analysis
Biodegradable material mechanics

1 Introduction

Percutaneous coronary intervention (PCI) with stenting has since its invention by Andreas Grüntzing in 1974 [1] grown to become the first choice to treat coronary artery stenosis, a disease caused by atherosclerosis, which is the local disposition of greasy plaques on the inner side of the arterial wall. PCI is a minimal invasive procedure and replaces the open-chest bypass surgeries, which has led to shorter revalidation times and decreased mortality rates. Conventional durable stents, including both bare metal stents (BMS) and drug eluting stents (DES), are made of metal alloys such as stainless steel, cobalt–chromium or platinum–chromium and remain permanently inside the human body after placement. However, the need for mechanical support for the healing artery is temporary and beyond the first few months there are potential disadvantages of a permanent metallic prosthesis: excessive neo-intimal tissue growth causes in-stent restenosis and prolonged exposure of the metallic stent surface to the blood stream increases the risk for late stent thrombosis [2]. Despite the fact that repeated

© Springer Nature Switzerland AG 2019
K. Arkusz et al. (Eds.): BIOMECHANICS 2018, AISC 831, pp. 81–89, 2019.
https://doi.org/10.1007/978-3-319-97286-2_7

percutaneous and surgical revascularization procedures due to restenosis could be reduced to 50–70% by the use of DES [3, 4], they still intrinsically display some unwanted features. Permanent stents are a chronic irritation to the host; they interfere with future cardiac interventions and do not match the natural vessel's behavior [5–7].

Abovementioned problems promise to be solved with the advent of fully biodegradable scaffolds (BRSs). This fairly new technology offers the possibility of transient scaffolding of the vessel to prevent acute vessel closure and recoil while transiently eluting an antiproliferative drug to counteract the constrictive remodeling and excessive neointimal hyperplasia and other advantages named in detail in reference [8].

Two classes of biomaterials are currently being used in biodegradable stent technology: biodegradable polymers and biocorrodible metal alloys. Polymers can be tailored to have a well-defined degradation pattern but have relatively poor mechanical properties. In contrast, biocorrodible metals such as magnesium alloys have excellent mechanical characteristics but display more complex and less predictive degradation behaviour. Metallic stents of any geometry generally have ample radial strength and the problems of enhancing axial flexibility and reducing neointimal hyperplasia have become the greater focus. Biodegradable polymers have a strength and stiffness that are between one and two orders of magnitude lower than those for modern stent metals. Therefore, radial strength is a great concern for biodegradable stents. Research on biodegradable stents has delved into the issue of the actual degradation, but the primary mechanical properties of the device have not been adequately addressed. Therefore developing biodegradable stents comprises challenges in making a stent that has sufficient radial strength for an appropriate duration, that does not have unduly thick struts, that can be a drug delivery vehicle and where degradation does not generate an unacceptable inflammatory response [9].

As it was mentioned, one of the concerns for polymeric BRSs is their mechanical performance, especially their interaction with blood vessels during and post deployment. Finite element method (FEM) has been particularly useful in understanding stent performance. However, the majority of FEM analyses of stent deployment, including many of the latest ones, were focused on the mechanical behaviour of metallic stents, i.e., expansion, dogboning, recoiling, design and stent–artery interaction (e.g. [10–12]). There is extremely limited work devoted to modelling the deformation of bioresorbable polymeric stents. Rare examples of such studies are reference [13] compared mechanical performance of PLLA stents for three different geometries with varying materials stiffness and [14] focusing on the development of an implicit finite element strategy, as opposed to an explicit one, to study the mechanical behaviour, such as stress distribution, recoiling an dogboning effects, of PLLA stents. Both works ignored the interaction with artery.

This paper focuses on the another aspect influencing polymeric BRSs mechanical behavior, which is operational temperature. It is well known, that materials such PLLA or PLGA are temperature sensitive [15]. Presented are FE analyses that include effects of process temperature in constitutive modeling.

2 FE Model of BRSs

A stent implant fabricated of bioresorbable polymer material was considered. The total length of the stent, l_{total}, was assumed to be equal to 8.2 mm, and the nominal inner diameter, $d_{nominal}$, was 3 mm. To deliver the stent into a stenosed region of the coronary artery, its outer diameter should be as small as possible. The initial outer diameter d_{base} of the stent is approximately 2 mm. During the manufacturing process, the stent is crimped on the balloon to an outer diameter d_{crimp} of approximately 1 mm using a specially designed crimping device [16]. Subsequently, the stent is expanded with a high-pressure balloon to a nominal inner diameter of 3 mm. To overcome the recoil phenomenon, strategies involving multiple inflations of the balloon might be needed.

First, a complex FE analysis of both the crimping and inflation process of the bioresorbable coronary stent (initial design) was performed. The complex analysis included the entire process of coronary stent crimping and inflation of the expandable folded balloon with minimal simplification. The results of the analysis were used as key parameters in the development of a simplified parametric FE model of the system. The complex model of stent is shown in Fig. 1.

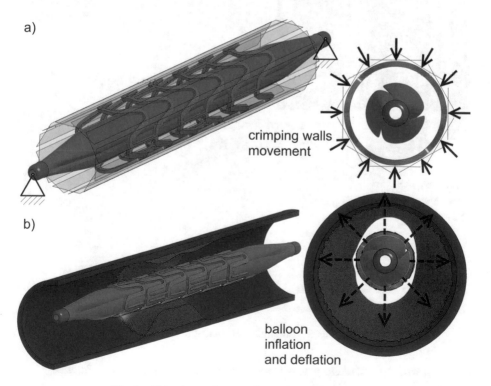

a)

crimping walls movement

b)

balloon inflation and deflation

Fig. 1. Crimping and expansion stage analysis set-up

2.1 Simplified Model of Stent

Based on the full model of stent the model of representative cut-out was developed. The procedure consisted of two steps: plain 2D FE model generation using the pre-processor script, followed by transformation to 3D using the Fortran script developed by authors. The FE model of cut-out is shown in Fig. 2. The model was made of 3D 8-node hexagonal elements. Balloon and artery were modelled using rigid surfaces that were connected to the nonlinear springs. Stiffness of springs were calculated based on the results of global stent model mentioned above and were representing stiffness of the balloon and blood vessel. The mechanism of contact between the surfaces of the balloon and the inner surface of the stent, as well as balloon and stent self-contact interfaces, was defined by a penalty formulation method, where FE code is placing normal interface springs between all nodes that penetrate the contact surface [17]. Nonlinear static analyses were adopted using Newton-Raphson incremental-iterative method. More information about the adopted solving algorithm as well as implemented convergence parameters to ensure the contact stability can be found in [17].

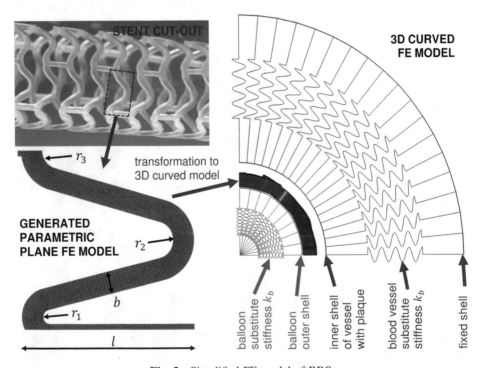

Fig. 2. Simplified FE model of BRSs

2.2 Constitutive Modeling and Consideration of Temperature Influence

Mechanical properties of the stent material were described by piecewise linear plasticity constitutive model. The data were taken from the experiments conducted in different ambient temperatures (see Fig. 3).

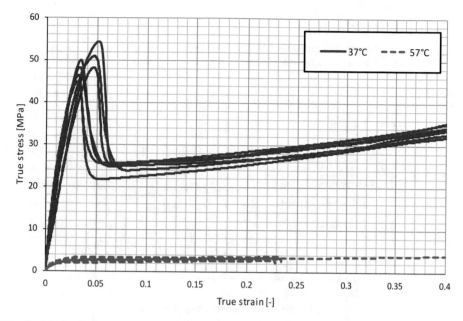

Fig. 3. Mechanical properties of the stent material in different temperatures (data acquired during experiments)

Influence of the ambient temperature on stent behaviour was investigated by defining three different simulations scenarios:

1. Crimping, expansion and deflation stages are carried out in the same temperature of 37 °C,
2. Crimping stage is conducted in the elevated temperature of 57 °C, while expansion and deflation are carried out in the temperature of 37 °C. Residual stresses left after crimping are canceled and expansions starts with none residual stresses in the material of stent.
3. As previously, crimping is conducted in the temperature of 57 °C and expansion, followed by deflation are done in 37 °C. This time, though, the residual stresses are included in the analysis and thus the expansion stage initiates with the strain field developed during first (crimping) stage.

3   Results

Radial strength has always been a primary design requirement for stents to be able to withstand compressive forces exerted by the vascular wall and prevent acute elastic recoil of the artery. Authors to assess the radial force implemented the procedure allowing to recalculate the contact pressure which is the function of the stiffness scaling factor, the stiffness modulus of contact segment, the scaling parameter, the character-istic length of FE elements and the determined penetration [17]. The adopted friction

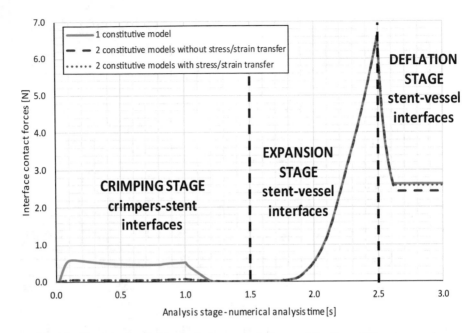

Fig. 4. Change of radial force during stent deployment for three different scenarios

model in the performed computations is a standard Coulomb friction law expressed in terms of the tangential contact stress. Figure 4 shows radial force developed by investigated BRSs in all three scenarios.

As it can be seen, different temperatures during crimping does not influence much radial force carried out by stent structure. The difference can be noticed only in the first stage of deployment and at the end of the process. At the first stage difference comes from the fact, that elevated temperature of 57 °C is responsible of reduction of polymer stiffness. At the end of the process radial force calculated in second scenario is smaller than in other two cases. This comes from the fact, that in second scenario, due to cancellation of the strain field after crimping, plastic strain developed after expansion are lower than in the other scenarios resulting in lower stiffness of the whole stent structure.

It should be noticed, that maximum force is equal in all three cases because it comes from the stiffness of the blood vessel expanded by balloon and not the stiffness of the stent structure.

Figure 5 shows history of maximum plastic strain at end of expansion for all three scenarios captured from the most strenuous areas (knee areas). The difference between scenario 2 (residual strain after crimping are not taken into account) and scenario 3 is clearly seen. As it was mentioned, lack of the residual strain results in lower value of the maximum strain causing drop of the overall stiffness of the stent.

Strain distributions for all three investigated cases are shown in Fig. 6. They are very similar to each other except different values. This shows, that overall structure behaviour remains the same in all analyses. It is worth noticed, that maximum values of

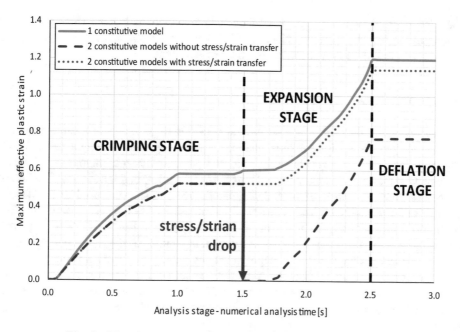

Fig. 5. Histories of maximum plastic strains for all three scenarios.

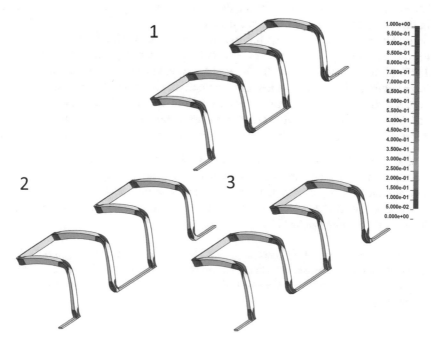

Fig. 6. Plastic strain distribution in stent structure for all three scenarios

the plastic strain in Fig. 6 are lower than values shown in Fig. 5. This is because averaging of discrete values provided by FE solver by postprocessor graphic program in order to present continuous maps of strain distribution.

4   Conclusions

The current study focused on the development of a new geometry of a cardiovascular stent fabricated from bioresorbable material considering three different temperature regimes. The complex analysis of the analysed pattern with minimum simplification provided detailed results on stent behaviour during the crimping, inflation and deflation processes (three scenarios of stent deployment). These data were used in the development of the simplified model. The simplified model is based on 3D elements for the stent and rigid or spring elements for both the balloon and the vessel. The proposed setup allowed to check whenever the elevated temperature during crimping influence stent behaviour during rest of the deployment procedure. The main advantage of the proposed methodology is also the wall time required for the computation which remains significantly shorter (approximately 20 times) in comparison with the same simulation using full 3D model with the implementation of integration scheme using explicit solution.

Obtained results can be summarized as follows:

- Elevation of the ambient temperature during crimping stage of deployment does not influence much the value of the radial force.
- Stent deployment is multistage process. It is vital for the simulation results to keep track of any changes of strain or deformation between deployment stages.
- The limit of safe deformation during implantation and further operation is reflected in maximum plastic strain after deployment. If the stent is crimped in elevated temperature and stress relaxation occurs before implantation, stent is capable to deform much more then stent crimped in lower temperature. This phenomena can be used to develop stent with higher radial strength by optimization its geometry.

Authors strongly believe that this work as well as the previous papers present great potential of implications for future studies of biomedical engineering in terms of numerical simulations and may help better understand the mechanisms governing the stent expansion, and, therefore, it may contribute to the development of new medical equipment and technologies [17].

Acknowledgements. The study was supported by the NCBiR within project "Apollo" (STRATEGMED) and the Interdisciplinary Centre for Mathematical and Computational Modelling (ICM) of the University of Warsaw under grant no. GB65-19. This support is gratefully acknowledged.

References

1. Barton, M., Grüntzig, J., Husmann, M., Rösch, J.: Balloon angioplasty – the legacy of andreas grüntzig, M.D. (1939–1985). Front. Cardiovasc. Med. **1,** 15 (2014)
2. Garg, S., Patrick, W.: Serruys, coronary stents: current status. J. Am. College Cardiol. **56** (10), 1–42 (2010)
3. Moses, J.W., Leon, M.B., Popma, J.J., Fitzgerald, P.J., Holmes, D.R., O'Shaughnessy, C., Caputo, R.P., Kereiakes, D.J., Williams, D.O., Teirstein, P.S., Jaeger, J.L.: Sirolimus-eluting stents versus standard stents inpatients with stenosis in a native coronary artery. New Engl. J. Med. **349**(14), 1315–1323 (2003)
4. Stone, W.G., Ellis, G.S., Cox, A.D., Hermiller, J., O'Shaughnessy, C., Mann, T.J., Turco, M., Caputo, R., Bergin, P., Greenberg, J., Popma, J.J., Rusell, E.M.: A polymer-based, paclitaxel-eluting stent in patients with coronary artery disease. N. Engl. J. Med. **350**(3), 221–231 (2004)
5. Waksman, R.: Promise and challenges of bioabsorbable stents. Cathet. Cardiovasc. Interv. **70**, 407–414 (2007)
6. Grech, E.D.: ABC of Interventional Cardiology, 2nd edn. Wiley-Blackwell, West Sussex (2011)
7. Erne, P., Schier, M., Resink, T.J.: The road to bioabsorbable stents: reaching clinical reality? Cardiovasc. Interv. Radiol. **29**, 11–16 (2006)
8. Onuma, Y., Serruys, W.P.: Bioresorbable scaffold the advent of a new era in percutaneous coronary and peripheral revascularization? New Drugs Technol. Circ. **123**(7), 779–797 (2011)
9. Ielasi, A., Latib, A., Colombo, A.: Current and future drug-eluting coronary stent technology. Expert Rev. Cardiovas. Ther. **9**, 485–503 (2011)
10. Chua, S.D., MacDonald, B., Hashmi, M.: Finite element simulation of stent and balloon interaction. J. Mater. Process. Technol. **143**, 591–597 (2003)
11. Lally, C., Dolan, F., Prendergast, P.: Cardiovascular stent design and vessel stresses: a finite element analysis. J. Biomech. **38**, 1574–1581 (2005)
12. Bukala, J., Malachowski, J., Kwiatkowski, P.: Finite element analysis of the percutaneous coronary intervention in a coronary bifurcation. Acta Bioeng. Biomech. **16**(4), 23–31 (2014)
13. Pauck, R.G., Reddy, B.D.: Computational analysis of the radial mechanical performance of PLLA coronary artery stents. Med. Eng. Phys. **37**, 7–12 (2014)
14. Debusschere, N., Segers, P., Dubruel, P., Verhegghe, B., DeBeule, M.: A finite element strategy to investigate the free expansion behavior of a biodegradable polymeric stent. J. Biomech. **48**, 2012–2018 (2015)
15. Ge, H., Yang, F., Hao, Y., Guangfeng, W., Zhang, H., Dong, L.: Thermal, mechanical, and rheological properties of plasticized poly(L-lactic acid). J. Appl. Polym. Sci. **127**(4), 2832–2839 (2013)
16. Bukala, J., Malachowski, J., Kwiatkowski, P.: Numerical analysis of stent expansion process in coronary artery stenosis with the use of non-compliant balloon. Biocybern. Biomed. Eng. **36**(1), 145–156 (2016)
17. Bukala, J., Kwiatkowski, P., Malachowski, J.: Numerical analysis of crimping and inflation process of balloon-expandable coronary stent using implicit solution. Int. J. Numer. Methods Biomed. Eng. **33**(12), 1–11 (2017). https://doi.org/10.1002/cnm.2890

Comparison of Methods for Computing a Target Point for Aspirations and Biopsies

Adam Ciszkiewicz$^{(\boxtimes)}$ ⓘ and Grzegorz Milewski

Institute of Applied Mechanics,
Cracow University of Technology, Cracow, Poland
adam.ciszkiewicz@gmail.com, milewski@mech.pk.edu.pl

Abstract. The aim of this study was to compare three methods for computing a target point for use in autonomous or semi-autonomous aspirations and biopsies. Given a 3D binary image of the object of interest, the procedures computed the target point. The following approaches were tested: the method #1 - center of mass, the method #2 - largest projection area + largest empty circle and the method #3 - largest empty circle + largest empty circle. Each procedure was tested on four cases obtained from Magnetic Resonance Imaging scans used to diagnose Baker's cysts. The methods were analyzed and compared in terms of their safety and computation time. In terms of safety, the best results were obtained with the third procedure, which used the largest empty circle + largest empty circle combination. The second method - the largest projection area + largest empty circle - offered good compromise between safety and computation time. It can be used to estimate target points for medical tool path planning in aspiration or biopsy.

Keywords: Largest empty circle · Center of mass · Voxel

1 Introduction

The world population is ageing rapidly, while the shortages in the health sector are quickly becoming significant [1]. This explains the increased interest in autonomous and semi-autonomous surgical systems as they can potentially offload surgeons. Such systems often require modules for: modeling soft tissues and joints [2–9], surgery planning and optimization [10–13], path planning for medical tools [14–16]. Before the path planning can be performed, the starting and the target tool location have to be specified. Some path planning approaches allow for no starting location and only an estimate of the target, nevertheless, typical methods work best with these two parameters given as the input.

In semi-autonomous systems, the target location can be selected by the user based on a set of Magnetic Resonance Imaging (MRI) scans. In this case, the user is responsible for selecting viable and safe target locations. With autonomous surgical systems, the problem is much more difficult. Firstly, the system has to diagnose the patient and, if needed, classify him for the surgical procedure [17, 18]. Then, the structure that requires treatment is segmented from the set of scans [19–22]. Finally, the target location has to be obtained based on the segmented image.

© Springer Nature Switzerland AG 2019
K. Arkusz et al. (Eds.): BIOMECHANICS 2018, AISC 831, pp. 90–97, 2019.
https://doi.org/10.1007/978-3-319-97286-2_8

The problem of computing viable target locations for the tool is of huge importance with regards to the safety of the surgical procedure. Therefore, the aim of this study was to compare three methods for computing a target point for use in autonomous or semi-autonomous aspirations and biopsies. Given a 3D, binary image of the object of interest, the procedures computed the target point. In total, three approaches were considered: the method #1 - center of mass, the method #2 - largest projection area + Largest Empty Circle (LEC) and the method #3 - LEC + LEC. Each one was tested on four cases obtained from actual Magnetic Resonance Imaging (MRI) scans used to diagnose Baker's cysts. Then, the procedures were compared in terms of their safety and computation time.

2 Method

2.1 Input to the Procedures

As mentioned before, the input to all of the considered procedures was a 3D binary image containing the segmented object of interest. In this study, the segmented objects represented Baker's cysts. The method presented in [23] was used to segment the cysts from spatial MRI images. Firstly, a point inside the cyst was selected using a custom graphical user interface. Then, the intensity of the image, in the vicinity of the selected point, was used to compute the lower and the upper limit for the thresholding. After the thresholding, spatial labeling was performed. Only the object, which contained the selected point, was left in the final image. In total, 4 cyst images were obtained from Short Tau Inversion Recovery (STIR) and Proton Density Weighted Spectral Attenuated Inversion Recovery (PDW SPAIR) MRI scans in the transverse plane. The images of varied quality, resolution, cyst location/size were imported into Python using Pydicom [24].

Each segmented image was stored in the computer memory as a 3D matrix. Its nonzero elements were the small cuboidal volumes (voxels) that belonged to the segmented object. Using the matrix, the procedures computed the target points. In this study, these points represented the final positions of the needle tip for Baker's cyst aspiration. As the input MRI scans were obtained in the transverse plane, it was assumed that the aspiration was to be performed in the transverse plane. Therefore, the safety of the procedure was also analyzed in the transverse plane. The considered methods for computing the points were explained in detail below.

2.2 The Considered Methods

The Method #1 - Center-of-Mass
In the first method, the target point was coincident with the center of mass of the object. The center of mass p_{cm} was computed as follows:

$$p_{cm} = \left[\sum_i^n x_i \quad \sum_i^n y_i \quad \sum_i^n z_i \right]^T \Big/ n, \tag{1}$$

where: p_{cm} - the center of mass of the object, x_i, y_i, z_i - the coordinates (indexes) of the nonzero elements in the input 3D image ($i = 1.. n$), n - the number of object's voxels (nonzero elements in the image). In this paper, the center of mass was computed using Scipy [25]. This approach didn't maximize the safety of the procedure. Furthermore, for highly concave shapes, the computed point may lie outside the boundary of the object.

The Method #2 and #3 - Based on the LEC
Since, the surgical procedure was assumed to take place in the transverse plane, the computation of the target point could be split into two subprocedures: the selection of the optimal insertion plane (parallel to the transverse plane) and the computation of the optimal target point within the selected plane. In this study, two criterions were considered for the plane selection. In the first one, the best plane corresponded to the plane with the largest area of the object's projection (the method #2). In the second one, the plane was selected based on the LEC radius inside the object's projection (the method #3). After the plane selection, the target point was computed as the center of the LEC in the selected plane for both methods. To compute the LEC, a method based on Voronoi diagrams was used [26].

3   Results and Discussion

The considered procedures were tested on 4 segmented Baker's cyst with varied size, location and shape. The obtained results were summarized in Table 1. The table includes: the coordinates of the obtained target points, the radii of the LEC based on these points and the run times of the procedures. In this study, the LEC radius was used as the measure of the safety of the method for computing the target points. Geometrically, this radius represents how far off the computed point the tool can be inserted without going outside of the cyst. It can also be seen as the maximum error margin for the tool. Note that the LEC for the method #1 was computed only to compare its safety to the other methods - it wasn't required otherwise. In more complex analyses, which take into account fluid dynamics, the obtained points could be used as a starting solution.

In terms of the computation time, the method #2 was the fastest. The run times varied between the cases, nevertheless, the method #2 was at least 3 times faster than the method #1 and 4 times faster than the method #3. As mentioned before, the procedures were implemented and tested in Python with popular numeric libraries (Numpy and Scipy). The run times may differ with different implementations or in lower-level languages, such as C++.

When considering the safety of the surgical procedure, the method #3 returned the best results with the largest LEC radius in all the considered cases (see Table 1). Nevertheless, the method #2 was only slightly worse, offering nearly the same safety with significantly better performance. On the other hand, the method #1 returned significantly worse results than #2 and #3. To study the results in detail, the obtained target points were visualized in Figs. 1, 2, 3 and 4 with the 3D images of the cysts.

Table 1. The summary of the obtained results, where: x, y, z - the coordinates of the obtained target points (dimensionless as they weren't scaled to the world coordinate system), r - the radius of the LEC based on the target point in the selected projection (dimensionless as it wasn't scaled to the world coordinate system), t - the computation time in [s].

CASE #1	x [-]	y [-]	z [-]	r [-]	t [s]
method #1	208	278	25	41.8	0.3
method #2	215	278	27	50.8	0.1
method #3	215	274	28	51.1	0.9
CASE #2	x [-]	y [-]	z [-]	r [-]	t [s]
method #1	165	207	21	26.3	0.3
method #2	166	226	22	32.5	0.1
method #3	166	224	23	33.8	0.4
CASE #3	x [-]	y [-]	z [-]	r [-]	t [s]
method #1	147	370	22	31.2	0.7
method #2	132	369	21	37.0	0.2
method #3	137	373	22	37.4	1.1
CASE #4	x [-]	y [-]	z [-]	r [-]	t [s]
method #1	259	497	25	20.9	0.6
method #2	247	484	26	32.0	0.2
method #3	247	484	26	32.0	1.1

In the case #1 (see Fig. 1), all procedures returned viable aspiration points. This was mostly due to the cyst being relatively large and convex. The safest point was computed with the method #3. Nevertheless, the difference between the method #2 and #3 was negligible in this regard.

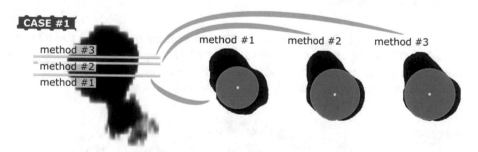

Fig. 1. The results from the case #1 - the target points obtained with the three methods along with a 3D image of the cyst.

As seen in Fig. 2, the method #3 returned the best results. The method #2 was again negligibly worse. The point computed by the method #1 was centered on the projection. Nevertheless, its LEC was of low radius when compared to the other methods.

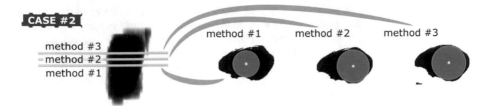

Fig. 2. The results from the case #2 - the target points obtained with the three methods along with a 3D image of the cyst.

The results obtained from the case #3 were interesting. The methods #2 and #3 returned the same plane, but with different target points within the plane. As seen in Table 1, the LEC based on the point from the method #3 was almost 20% larger than that from the method #1. The method #2 returned a different plane than #3, but with comparable safety.

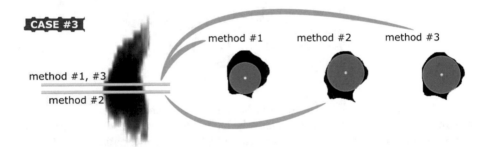

Fig. 3. The results from the case #3 - the target points obtained with the three methods along with a 3D image of the cyst.

In the case #4 (see Fig. 4), the points obtained with the methods #2 and #3 were coincident. The point obtained with the method #1 was significantly less safe when compared to points from methods #2 and #3. Furthermore, the computed insertion plane was near the middle of the cyst for all the methods. It is worth noting that the cyst

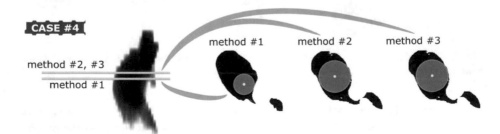

Fig. 4. The results from the case #4 - the target points obtained with the three methods along with a 3D image of the cyst.

projections had multiple disconnected regions. All the methods handled this scenario correctly. Nevertheless, the method #1 isn't guaranteed to return a point inside the object in such cases.

To summarize, the best results, in terms of safety, were returned by the method #3 - LEC + LEC. The method #2 - largest projection area + LEC - was the fastest. In all of the considered cases, the method #2 offered nearly the same safety as the method #3. The low numerical complexity and the high safety make it a good choice for computing target points for use in path planning procedures. The method #1 - center of mass - returned unsafe target points and shouldn't be employed in software for surgical path planning.

When comparing the presented methods to the ones available in the literature, it is worth noting that most authors assume that the target point is given by the user [14, 27, 28], which makes the system semi-autonomous. The method #1 has been applied previously in biopsies [29]. Nevertheless, as shown by our results, the methods based on LEC provide much safer targets. It is also worth mentioning that some of the path planning approaches allow for estimating the final location along with the path in the path planning procedure [10].

As mentioned before, the proposed methods for computing the target point were tested specifically for the aspiration of Baker's cysts in the transverse plane. Nevertheless, the approach is general and can be used with any segmented objects for computing safe and viable target points for surgery.

4   Conclusion

In this study, three methods for computing a target point for use in systems for autonomous or semi-autonomous surgery were compared. The following approaches were considered: method #1 - center of mass, method #2 - largest projection area + LEC and the method #3 - LEC + LEC. Each procedure was tested on four cases obtained from MRI scans used to diagnose Baker's cysts. Firstly, the scans were segmented with thresholding and labeling. Then, the segmented cysts were inputted into the tested methods. The methods were analyzed and compared in terms of their safety and computation time. The safety was measured using the radius of the LEC based on the computed point. With this criterion, the third method - LEC + LEC - was the safest. The second method - largest projection area + LEC - offered the best performance with nearly identical safety to that of the method #3 - LEC + LEC. It provided a good compromise between the computation time and the safety. Therefore, it can be used to estimate target points for medical tool path planning in aspiration or biopsy.

References

1. Scheil-Adlung, X., Behrendt, T., Wong, L.: Health sector employment: a tracer indicator for universal health coverage in national Social Protection Floors. Hum. Resour. Health **13**, 1–8 (2015). https://doi.org/10.1186/s12960-015-0056-9
2. Ciszkiewicz, A., Milewski, G.: A novel kinematic model for a functional spinal unit and a lumbar spine. Acta Bioeng. Biomech. **18** (2016). https://doi.org/10.5277/abb-00324-2015-03
3. Sancisi, N., Parenti-Castelli, V.: A 1-Dof parallel spherical wrist for the modelling of the knee passive motion. Mech. Mach. Theory **45**, 658–665 (2010). https://doi.org/10.1016/j.mechmachtheory.2009.11.009
4. Pappalardo, O.A., Sturla, F., Onorati, F., Puppini, G., Selmi, M., Luciani, G.B., Faggian, G., Redaelli, A., Votta, E.: Mass-spring models for the simulation of mitral valve function: looking for a trade-off between reliability and time-efficiency. Med. Eng. Phys. **47**, 93–104 (2017). https://doi.org/10.1016/j.medengphy.2017.07.001
5. Basafa, E., Farahmand, F.: Real-time simulation of the nonlinear visco-elastic deformations of soft tissues. Int. J. Comput. Assist. Radiol. Surg. **6**, 297–307 (2011). https://doi.org/10.1007/s11548-010-0508-6
6. Cicek, Y., Duysak, A.: The modelling of interactions between organs and medical tools: a volumetric mass-spring chain algorithm. Comput. Methods Biomech. Biomed. Eng. **17**, 488–496 (2014). https://doi.org/10.1080/10255842.2012.694875
7. Assi, K.C., Grenier, S., Parent, S., Labelle, H., Cheriet, F.: A physically based trunk soft tissue modeling for scoliosis surgery planning systems. Comput. Med. Imag. Graph. **40**, 217–228 (2015). https://doi.org/10.1016/j.compmedimag.2014.11.002
8. Mohammadi, A., Ahmadian, A., Rabbani, S., Fattahi, E., Shirani, S.: A combined registration and finite element analysis method for fast estimation of intraoperative brain shift; phantom and animal model study. Int. J. Med. Robot. Comput. Assist. Surg. **13**, e1792 (2017). https://doi.org/10.1002/rcs.1792
9. Villard, P.-F., Hammer, P.E., Perrin, D.P., del Nido, P.J., Howe, R.D.: Fast image-based mitral valve simulation from individualized geometry. Int. J. Med. Robot. Comput. Assist. Surg. **14**, e1880 (2018). https://doi.org/10.1002/rcs.1880
10. Ciszkiewicz, A., Milewski, G.: Path planning for minimally-invasive knee surgery using a hybrid optimization procedure. Comput. Methods Biomech. Biomed. Eng. **21**, 47–54 (2018). https://doi.org/10.1080/10255842.2017.1423289
11. Olszewski, R., Villamil, M.B., Trevisan, D.G., Nedel, L.P., Freitas, C.M.D.S., Reychler, H., Macq, B.: Towards an integrated system for planning and assisting maxillofacial orthognathic surgery. Comput. Methods Programs Biomed. **91**, 13–21 (2008). https://doi.org/10.1016/j.cmpb.2008.02.007
12. Luboz, V., Ambard, D., Swider, P., Boutault, F., Payan, Y.: Computer assisted planning and orbital surgery: patient-related prediction of osteotomy size in proptosis reduction. Clin. Biomech. **20**, 900–905 (2005). https://doi.org/10.1016/j.clinbiomech.2005.05.017
13. Adolphs, N., Haberl, E.-J., Liu, W., Keeve, E., Menneking, H., Hoffmeister, B.: Virtual planning for craniomaxillofacial surgery–7 years of experience. J. Craniomaxillofac. Surg. **42**, e289–e295 (2014). https://doi.org/10.1016/j.jcms.2013.10.008
14. Napalkova, L., Rozenblit, J.W., Hwang, G., Hamilton, A.J., Suantak, L.: An optimal motion planning method for computer-assisted surgical training. Appl. Soft Comput. **24**, 889–899 (2014). https://doi.org/10.1016/j.asoc.2014.08.054
15. Vrooijink, G.J., Abayazid, M., Patil, S., Alterovitz, R., Misra, S.: Needle path planning and steering in a three-dimensional non-static environment using two-dimensional ultrasound images. Int. J. Rob. Res. **33**, 1361–1374 (2014). https://doi.org/10.1177/0278364914526627

16. Hassouna, M.S., Farag, A.A., Hushek, S.G.: 3D path planning for virtual endoscopy. Int. Congr. Ser. **1281**, 115–120 (2005). https://doi.org/10.1016/j.ics.2005.03.142
17. Zarychta, P., Badura, P., Pietka, E.: Comparative analysis of selected classifiers in posterior cruciate ligaments computer aided diagnosis. Bull. Polish Acad. Sci. Tech. Sci. **65**, 63–70 (2017). https://doi.org/10.1515/bpasts-2017-0008
18. Szwarc, P., Kawa, J., Rudzki, M., Pietka, E.: Automatic brain tumour detection and neovasculature assessment with multiseries MRI analysis. Comput. Med. Imag. Graph. **46**, 178–190 (2015). https://doi.org/10.1016/j.compmedimag.2015.06.002
19. Shan, L., Zach, C., Charles, C., Niethammer, M.: Automatic atlas-based three-label cartilage segmentation from MR knee images. Med. Image Anal. **18**, 1233–1246 (2014). https://doi.org/10.1016/j.media.2014.05.008
20. Araújo, T., Abayazid, M., Rutten, M.J.C.M., Misra, S.: Segmentation and three-dimensional reconstruction of lesions using the automated breast volume scanner (ABVS). Int. J. Med. Robot. Comput. Assist. Surg. (2016). https://doi.org/10.1002/rcs.1767
21. Abdolali, F., Zoroofi, R.A., Otake, Y., Sato, Y.: Automatic segmentation of maxillofacial cysts in cone beam CT images. Comput. Biol. Med. **72**, 108–119 (2016). https://doi.org/10.1016/j.compbiomed.2016.03.014
22. Markiewicz, T., Dziekiewicz, M., Osowski, S., Maruszynski, M., Kozlowski, W., Boguslawska-Walecka, R.: Thresholding techniques for segmentation of atherosclerotic plaque and lumen areas in vascular arteries. Bull. Polish Acad. Sci. Tech. Sci. **63**, 269–280 (2015). https://doi.org/10.1515/bpasts-2015-0031
23. Ciszkiewicz, A., Lorkowski, J., Milewski, G.: A novel planning solution for semi-autonomous aspiration of Baker's cysts. Int. J. Med. Robot. Comput. Assist. Surg. **14**, e1882 (2018). https://doi.org/10.1002/rcs.1882
24. Mason, D.: SU-E-T-33: Pydicom: An Open Source DICOM Library. Med. Phys. **38**, 3493 (2011). https://doi.org/10.1118/1.3611983
25. van der Walt, S., Colbert, S.C., Varoquaux, G.: The NumPy array: a structure for efficient numerical computation. Comput. Sci. Eng. **13**, 22–30 (2011). https://doi.org/10.1109/MCSE.2011.37
26. Toussaint, G.T.: Computing largest empty circles with location constraints. Int. J. Comput. Inf. Sci. **12**, 347–358 (1983). https://doi.org/10.1007/BF01008046
27. Moustris, G.P., Hiridis, S.C., Deliparaschos, K.M., Konstantinidis, K.M.: Evolution of autonomous and semi-autonomous robotic surgical systems: a review of the literature. Int. J. Med. Robot. Comput. Assist. Surg. **7**, 375–392 (2011). https://doi.org/10.1002/rcs.408
28. Herghelegiu, P.-C., Manta, V., Perin, R., Bruckner, S., Gröller, E.: Biopsy planner–visual analysis for needle pathway planning in deep seated brain tumor biopsy. In: Computer Graphics Forum, pp. 1085–1094. Wiley Online Library (2012)
29. Liang, K., Rogers, A.J., Light, E.D., von Allmen, D., Smith, S.W.: 3D ultrasound guidance of autonomous robotic breast biopsy: feasibility study. Ultrasound Med. Biol. **36**, 173–177 (2010). https://doi.org/10.1016/j.ultrasmedbio.2009.08.014

Predictive Models in Biomechanics

John Rasmussen[(⊠)]

Aalborg University, Aalborg, Denmark
jr@mp.aau.dk

Abstract. This paper investigates the opportunity of predictive musculoskeletal models that do not require experimental input of kinematics and ground reaction forces. First, the requirements of such models are reviewed and, subsequently, an example model of running is derived by means of principal component analysis. The generation of different running styles using the model is demonstrated, and we conclude that this type of models has the potential to predict motion behavior given shallow input describing the individual.

Keywords: Musculoskeletal models · Running · Statistics
Principal component analysis

1 Background and Motivation

Musculoskeletal models based on inverse dynamics serve a variety of purposes, such as automotive development [1], orthopedics [2], exoskeleton design [3], aerospace applications [4], work place ergonomics [5] and development of sports equipment [6]. However, models of gait or other types of ambulation based on experimental data, i.e. motion capture data and force platforms, are so common that they have defined the field for several years. Although gait labs with the aforementioned equipment exist in many hospitals and research centers, musculoskeletal modelling has still not managed a breakthrough in clinical practice. The reasons may be cumbersome modelling processes and lack of validated models, but comparison with the emergence of parallel technologies in other fields reveals another possible reason, which we shall address in this paper.

1.1 CAE

Computer-Aided Engineering (CAE), primarily the finite element method, has had such a profound impact on the industrial sector that most modern, high-tech products would not have existed in the absence of this technology. It is inconceivable to design a modern aircraft, a car or a mobile telephone in the absence of computer simulation of the relevant physical processes. CAE is used for product development because it enables prediction of product behavior – the flow of air around the fuselage of an aircraft, the vibrational response of a car body, or the transfer of heat from the CPU of a mobile telephone – before the products have been materialized. The models predict the behavior for given design parameters, and the principal value of the models if bound to this predictive power.

© Springer Nature Switzerland AG 2019
K. Arkusz et al. (Eds.): BIOMECHANICS 2018, AISC 831, pp. 98–106, 2019.
https://doi.org/10.1007/978-3-319-97286-2_9

It is easily understood that predictive models cannot rely on experimental data. Models replicating experimental data are by definition retrospective because the experiment has to be conducted before the model can be analyzed. As mentioned above, musculoskeletal models often require motion capture data and recorded ground reaction forces as input, and such models are therefore retrospective.

Retrospective models are not without value; they can explain observed phenomena and add causality to empirical results. For instance, given two gait trials recorded with motion capture technology, force platforms and EMG equipment, a retrospective model may explain why a larger activation of the gastrocnemius muscle appears in one trial versus the other; the model provides additional insight that cannot be obtained from the experiment alone.

For musculoskeletal modelling in a wider context of development, product design, and clinical applications, models are only useful if they predict events and consequences in the future: Will one or another design of a wheelchair be less prone to causing shoulder overuse injuries? Will a given training regimen improve a high jumper's performance? Will a possible surgical procedure improve a disabled patient's capacity for ambulation? This type of models are predictive and, although they rely on much the same technology as retrospective models, they are fundamentally different. This paper explores the requirements and opportunities of such models and uses the case of running as an example.

1.2 Predictive Model Requirements

In inverse dynamics, which is the paradigm addressed in this paper, motion capture data and recorded ground reaction forces are input to the analysis. In predictive models, they cannot be derived from prior experiments, so these data must be converted from being input to being output.

Another, possibly more subtle, difference between retrospective and predictive models is that the former by definition relates to specific subjects from whom the data were collected. On the contrary, in many predictive applications, such as industrial product design, products are not tailor-made for a particular individual but for a cross section of the population, for instance all people using sports shoes of European size 42. This indicates that predictive models should have a statistical capacity to them, which obviously concerns the anthropometry of the model. However, it is easily concluded that anthropometry can also influence the kinematics and kinetics of the model; subjects with long legs move differently from subjects with short legs, and the body mass has a direct influence of the ground reaction forces. It is therefore obvious that statistics will influence most or all of the model parameters.

2 Methods

The following is entirely based on the AnyBody Modeling System [7], but the principles should be applicable to any other inverse dynamics-based musculoskeletal analysis system, for instance [8].

2.1   Prediction of Ground Reaction Forces

While measured ground reaction forces have been the standard in inverse dynamics simulations, they are not strictly necessary. For a subject standing statically on the ground, the ground reaction forces stem from gravity alone and are easily computed from static equilibrium. Moving models are more complicated, but interface forces can still be derived from the Newton-Euler equations of motion, given known motions and mass properties of the segments in the model. It has already been shown [9, 10] that these predictions can be surprisingly accurate for a variety of movements.

The implementation of ground reaction force prediction requires some form of contact analysis, which in numerical mechanics is complicated by three properties:

1. Contact forces can only be exchanged if there is indeed contact between the surfaces. This creates a highly nonlinear and even non-smooth dependency between kinematics and kinetics in the model.
2. Contact forces are unilateral, i.e. contacting surfaces can exchange compression but not tension. This furthermore adds non-smooth properties to the kinetics problem.
3. Contact force is always accompanied by friction, which in the simplest (Coulomb) form is limited by the compression between the surfaces and the friction coefficient.

For these reasons, contact models range among the more complex problems in mechanics. However, as we shall demonstrate in the following, the mathematical structure of contact mechanics problems is almost precisely equivalent to muscle recruitment [11], which enables an elegant contact solution in inverse dynamics problems.

The dynamic equilibrium equations described in detail in [7] take the form:

$$\mathbf{Cf} = \mathbf{r} \tag{1}$$

where $\mathbf{f}$ is a vector of internal forces, i.e. muscle forces and joint reactions, and $\mathbf{r}$ is a vector of externally applied forces. When force platform data are used, they become part of $\mathbf{r}$.

The internal forces, $\mathbf{f}$, fall in different categories. Traditionally, they are divided into muscle forces, $\mathbf{f}^{(M)}$, and joint reactions, $\mathbf{f}^{(R)}$, where the former are unilateral (muscles can only pull) and limited in strength by a muscle model, and the latter are bilateral and unlimited in strength:

$$\mathbf{C}\left\{ \begin{array}{c} \mathbf{f}^{(M)} \\ \mathbf{f}^{(R)} \end{array} \right\} = \mathbf{r} \tag{2}$$

$$f_j^{(M)} \geq 0 \, for \, j = 1..n_M \tag{3}$$

where n_M is the number of muscles in the system.

In all practical cases, Eqs. (2)–(3) are redundant, i.e. have infinitely many solutions. It is therefore a usual assumption that muscles are recruited according to an optimality criterion, such that the problem to be solved takes the form:

Minimize

$$\sum_{j=1}^{n_M} \left(\frac{f_j^{(M)}}{N_j}\right)^p \tag{4}$$

Subject to

$$\mathbf{C}\left\{\begin{array}{c} \mathbf{f}^{(M)} \\ \mathbf{f}^{(R)} \end{array}\right\} = \mathbf{r} \tag{5}$$

$$f_j^{(M)} \geq 0 \, for \, j = 1..n_M \tag{6}$$

where N_j is the strength of each muscle, i.e. the muscle model, and p is a power, $1 < p < \infty$. Please notice that N_j can be a function of muscle kinematics, allowing for implementation of Hill-type muscle models. For details about the formulation, properties and solution of this problem, please refer to [11].

It turns out that Eqs. (4)–(6) lend themselves to formulation of contact mechanics problems as well, if they are further detailed:

Minimize

$$\sum_{j=1}^{n_M} \left(\frac{f_j^{(M)}}{N_j}\right)^p + \sum_{j=1}^{n_C} \left(\frac{f_j^{(C)}}{K_j}\right)^p \tag{7}$$

Subject to

$$\mathbf{C}\left\{\begin{array}{c} \mathbf{f}^{(M)} \\ \mathbf{f}^{(C)} \\ \mathbf{f}^{(R)} \end{array}\right\} = \mathbf{r} \tag{8}$$

$$f_j^{(M)} \geq 0 \, for \, j = 1..n_M \tag{9}$$

$$f_j^{(C)} \leq 0 \, for \, j = 1..n_C \tag{10}$$

where $f^{(C)}$ is the vector of n_C contact forces in the model, and K_j is the strength of the j'th contact force. Please notice that the mathematical structure of Eqs. (7)–(10) is identical to the structure of Eqs. (4)–(6) and therefore can be solved using the same numerical algorithms. Since strengths such as N_j and K_j can depend on kinematics, we can implement the rule:

$$K_j \begin{cases} \gg N_j \text{ when proximal to contact surface} \\ = 0 \quad \text{otherwise} \end{cases} \tag{11}$$

where the first condition reflects the fact that most supporting surfaces such as floors are strong compared with the humans resting on them, and the second condition ensures inactivity of the contact force when the body is remote from the contact surface, for instance in the swing phase of the gait cycle. The unilateral nature of contact is assured by Eq. (10). It is further possible to ensure Coulomb friction as a linear inequality constraint between normal contact forces and additional tangential friction forces.

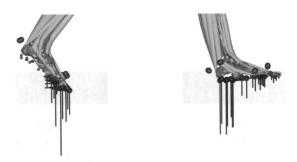

Fig. 1. Simulated ground reaction forces in a gait model.

Figure 1 shows a typical case of simulated contact forces distributions under the feet of a gait model. Please notice the presence of friction forces as well as normal forces, and please notice that the multiple contact points cause the ground reaction force center to move from the heel to the toe over the stance phase.

2.2   Statistical Capability of Models and Synthesis of Motion

Human motions are obviously very versatile and complex, and it is not within the reach of current technology to synthesize valid motions for abstract purposes for individual people. However, many important activities of daily living are recognizable, repeatable and cyclic. This includes ambulation, i.e. walking, running, hopping, stair walking and such. Furthermore, we can observe a dependence between anthropometry and motion characteristics; people with long legs are prone to relatively long steps, for instance. If such dependency applies to many different parameters, they could be exploited statistically by means of machine learning techniques, such as cluster analysis or principal component analysis (PCA), as will be demonstrated in the following.

We take the approach of compiling a number, m, of full-body motion capture trials of running. By means of musculoskeletal analysis with the AnyBody Modeling System [7] and parameter identification [12], each runner's anthropometry is determined and the motion is referred to anatomical joint angle variations. The latter will be cyclic in nature and are therefore easily approximated by Fourier series with a finite number of terms. The necessary number of terms varies somewhat depending on the complexity of each anatomical joint angle's variation. In the presented case, most joint angles required 3–5 terms in the Fourier series, and a few up to 10. Figure 2 illustrates the fitting of a typical joint angle function.

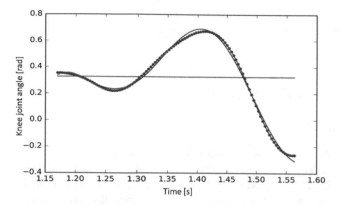

Fig. 2. Joint angle fitting. The dotted curve stems directly from the processed motion capture data. Three solid curves of red, magenta and cyan are hardly distinguishable and represent three, four and five terms respectively in the Fourier series.

Each joint angle movement is now described by a few Fourier coefficients. Anthropometric dimensions and Fourier coefficients of $m = 90$ running trials are stored as rows of length $n \approx 250$ in a matrix and subsequently subjected to PCA. The entire variation of running styles and anthropometry can subsequently be described by n principal components and ordered after decreasing variance as illustrated in Fig. 3.

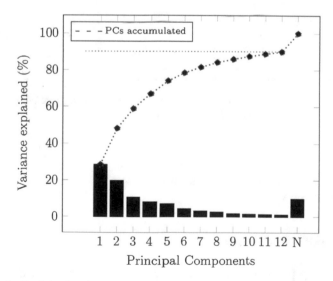

Fig. 3. Explained variance of principal components of the 90 running trials.

It turns out that 90% of the total variance can be described by the first 12 principal components. These components become parameters of a generic running model, and more or less components can be included in the parameter space, depending on the desired resolution in terms of running styles and anthropometry. We subsequently investigate the influence of the first two principal components on the resulting running pattern.

3   Results and Discussion

3.1   Kinematics

The average running pattern, corresponding to a zero value of all principal components, visually appears to be an archetypical medium speed running. The first principal component causes the model to jog slowly when varied two standard deviations in one direction and to sprint with long steps and large heel lift when varied oppositely, i.e. this component seems to be primarily associated with the running speed. The second principal component varies the model between a relaxed running style with little elbow and shoulder articulation and relatively low elevation of the heel in the forward swing phase, and an energetic style with large elbow and shoulder flexion, i.e. more pull with the arms. Figure 4 shows characteristic frames of the artificially generated running styles.

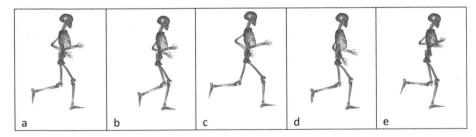

Fig. 4. Artificially generated running patterns. Average running (a), first PC − 2stdv (b), first PC + 2stdv (c), second PC − 2stdv (d), second PC + 2stdv (e).

3.2   Kinetics

Evaluation of the kinetics of the different running styles is a work in progress. No results are available yet.

4   Conclusions and Outlook

The kinematically plausible running styles generated from the principal components indicate that it is feasible to generate a parametric running model from empirical data in combination with PCA. However, 90 running trials is not much, considering that the data set has 250 parameters. Although the PCA reveals that many of the parameters are strongly correlated, a reliable model should be trained with many more running trials.

The running trials in this investigation were compiled from a variety of sources and were quite heterogeneous in terms of experimental protocols, so it is possible that some of the observed variance can be due to different experimental conditions. A systematic collection of a large number of running trials under similar experimental circumstances would improve the credibility of the results.

Despite these reservations, the results are encouraging. If, indeed, a reliable parametric running model can be obtained, then it is possible that we can identify – or predict – running patterns of individuals based on a shallow amount of kinematic data coupled with anthropometric information that is simple to obtain. It may also be possible, using the kinetic output, to investigate the influence of certain running style modifications on joint and muscle forces, thus identifying interventions that can reduce overuse injury risks or allow individuals with joint pain to continue exercising.

Acknowledgements. This work was supported by Innovation Fund Denmark.

References

1. Van Houcke, J., Schouten, A., Steenackers, G., Vandermeulen, D., Pattyn, C., Audenaert, E.A.: Computer-based estimation of the hip joint reaction force and hip flexion angle in three different sitting configurations. Appl. Ergon. **63**, 99–105 (2017)
2. Dell'Isola, A., Smith, S.L., Andersen, M.S., Steultjens, M.: Knee internal contact force in a varus malaligned phenotype in knee osteoarthritis (KOA). Osteoarthr. Cartil. **25**(12), 2007–2013 (2017)
3. Zhou, L., Li, Y., Bai, S.: A human-centered design optimization approach for robotic exoskeletons through biomechanical simulation. Robot. Auton. Syst. **91**, 337–347 (2017)
4. Lindenroth, L., Caplan, N., Debuse, D., Salomoni, S.E., Evetts, S., Weber, T.: A novel approach to activate deep spinal muscles in space—results of a biomechanical model. Acta Astronaut. **116**, 202–210 (2015)
5. Møller, S.P., et al.: Risk of subacromial shoulder disorder in airport baggage handlers: combining duration and intensity of musculoskeletal shoulder loads. Ergonomics **61**(4), 1–29 (2017)
6. Lee, H., Jung, M., Lee, K.K., Lee, S.H.: A 3D human-machine integrated design and analysis framework for squat exercises with a Smith machine. Sensors **17**(2), 299 (2017)
7. Damsgaard, M., Rasmussen, J., Christensen, S.T., Surma, E., De Zee, M.: Analysis of musculoskeletal systems in the AnyBody Modeling System. Simul. Model. Pract. Theory **14**(8), 1100–1111 (2006)
8. Delp, S.L., et al.: OpenSim: open-source software to create and analyze dynamic simulations of movement. IEEE Trans. Biomed. Eng. **54**(11), 1940–1950 (2007)
9. Skals, S., Jung, M.K., Damsgaard, M., Andersen, M.S.: Prediction of ground reaction forces and moments during sports-related movements. Multibody Syst. Dyn. **39**(3), 175–195 (2017)
10. Jung, Y., Koo, Y.J., Koo, S.: Simultaneous estimation of ground reaction force and knee contact force during walking and squatting. Int. J. Precis. Eng. Manuf. **18**(9), 1263–1268 (2017)

11. Rasmussen, J., Damsgaard, M., Voigt, M.: Muscle recruitment by the min/max criterion – a comparative numerical study. J. Biomech. **34**(3), 409–415 (2001)
12. Andersen, M.S., Damsgaard, M., MacWilliams, B., Rasmussen, J.: A computationally efficient optimisation-based method for parameter identification of kinematically determinate and over-determinate biomechanical systems. Comput. Methods Biomech. Biomed. Eng. **13** (2), 171–183 (2010)

Automatic Processing and Analysis of the Quality Healing of Derma Injury

Elena Semenova, Oleg Gerasimov, Elizaveta Koroleva,
Nafis Ahmetov, Tatyana Baltina⊙, and Oskar Sachenkov(✉)⊙

Kazan Federal University,
Kremlevskaya str. 18, 420000 Kazan, Russia
4works@bk.ru

Abstract. Automation of analyzing the biological data can increase the quality of analyses and decrease spending time. Analyze of the microscope's bitmaps is usual task in biology. To illustrate the proposed method we used analyzing collagen in dermis snapshots. Methodic to automatic analyses of microscope snapshots is presented. Object of analysis can be determine by color vector. Then the snapshot can be binarized and meshed. For every element we can restore distribution of the mean intercept length. Orientation of the objects can be calculated using approximation of the mean intercept length. Equation to estimate the quality of collagen recovery was presented. We used the method on samples of three types: no ficin group (N), ficin group (F), immobilized ficin (Fi). We tested 10 bitmaps for every group and we got results for all bitmaps according described technique. Quality of collagen recovery values was: for N group – 48% ± 8%, for F group 78 ± 7%, for Fi group 68 ± 9%. It can be concluded that ficin positively influence on dermas recovery. Received results are consistent with published results.

Keywords: Snapshot analysis · Fabric tensor · Collagen

1 Introduction

Automation of analyzing the biological data can increase the quality of analyses and decrease spending time. Analyze of the microscope's bitmaps is usual task in biology. In many cases there are typical problems to solve: count number of some biological object, analyze orientation of some biological object in snapshot. In paper we introduce method to analyze density and orientation of some biological object. To illustrate the proposed method we used analyzing collagen in dermis snapshots.

Let's introduce biological problem. Bacterial cells in biofilms are extremely resistant to drug treatment and to attacks from the immune system, which leads to chronic recurrent infections [1–3]. Many opportunistic bacteria (i.e. Staphylococcus, Micrococcus, Klebsiella, Pseudomonas, etc.) form biofilms on chronic and acute dermal wounds, preventing their cure and causing reinfection and sepsis [1, 3, 4]. Degradation of the biofilm matrix backbone, for example via enzymatic lysis, is an advantageous approach for controlling the growth and development of biofilms [5]. It was shown [6, 7] that protolytic enzymes are one of the most effective enzymes in the

© Springer Nature Switzerland AG 2019
K. Arkusz et al. (Eds.): BIOMECHANICS 2018, AISC 831, pp. 107–113, 2019.
https://doi.org/10.1007/978-3-319-97286-2_10

destruction of biofilm matrix by hydrolyzing both matrix proteins and adhesins (proteins that ensure the attachment of cells to hard surfaces and to other bacteria), and for cleavage of signal peptides of intercellular communication of Gram-positive bacteria [6], especially good for this purpose plant proteolytic enzyme ficin [8].

To estimate the quality of derma's recovery we can analyze the collagens orientation. It was shown, that in normal case collagens orientation is chaotic. While in case of injured derma the collagen fibers aligns in direction of maximum stress [9]. In this case we can estimate the quality of healing using relative density and orientation of the collagen fibers [10–12]. To calculate the orientation of the collagen fabric tensor was used [13–15].

2 Materials and Methods

2.1 Mean Intercept Length

Collagen distribution in area can be analyzed as an orthotropic media. To estimate main direction of collagen orientation we decided to use mean intercept length (MIL) [13, 16].

The MIL in a material can be calculated as the average distance, measured along a straight line $\vec{n}$, between two components of the material. In our case two components of the material are collagen and not collagen. MIL can be approximated by quadratic form [16], which can be described by the relationship:

$$L^{-2}(\vec{n}) = \vec{n} \cdot \tilde{M} \cdot \vec{n}, \tag{1}$$

where n is the unit vector in the direction of the mean intercept length measurement and length $L(\vec{n})$ is the mean intercept length.

In this case the problem of defining the orientation of collagen equivalent to eigenvalue and eigenvectors problem for tensor M. We introduce degree of anisotropy, it can be calculated as aspect ratio of the eigenvalues:

$$r = \frac{\lambda_2}{\lambda_1}, \tag{2}$$

where λ_1 - 1st eigenvalue (the largest one), λ_2 - 1st eigenvalue (the lowest one).

If the aspect ratio equal to one it means that there are no certain orientation for collagen, it's distribution is chaotic. Eigenvector related to the 1st eigenvalue (in case of aspect ratio lower than one) describes direction of the collagen's elongation.

2.2 Approximation

The MIL in a material can be stored as vector:

$$\vec{x}_i = (x_i, y_i); \ i = \overline{1, n} \tag{3}$$

In this case we can present quadratic form in general case as:

$$f(\vec{x}) = \vec{x} \cdot A \cdot \vec{x}^T + \vec{x} \cdot B + C = 0, \tag{4}$$

where $A = \begin{pmatrix} A_{11} & 2A_{12} \\ 2A_{12} & A_{22} \end{pmatrix}, B = \begin{pmatrix} B_1 \\ B_2 \end{pmatrix}$

The simplest non-geometric fit in this case is the one minimizing:

$$J(\vec{x}) = \sum_{i=1}^{n} (f(\vec{x}))^2 \rightarrow \min \tag{5}$$

To justify this method it can be noted that $f(\vec{x})$ equivalent to zero if and only if the points $\vec{x}$ lies on the curve, and $J(\vec{x})$ is small when the point lies near the curve. We introduce the vector of unknown parameters of the quadratic form (4):

$$\vec{\eta} = (A_{11}, A_{22}, A_{12}, B_1, B_2, C)^T \tag{6}$$

In this case the problem of the minimization (5) of the $J(\vec{x})$ can be presented:

$$J(\vec{x}, \vec{\eta}) = \sum_{i=1}^{n} (f(\vec{x}, \vec{\eta}))^2 \rightarrow \min \tag{7}$$

That's lead to:

$$\frac{\partial J(\vec{x}, \vec{\eta})}{\partial \eta_i} = 0 \ (i = 1, 6) \tag{8}$$

This problem is equal to system of linear equations [17]:

$$S \cdot \vec{\eta} = \vec{b} \tag{9}$$

But the matrix S is singular. Usually to solve this problem some additional equations added to system of linear equations (i.e. absolute value of vector $\vec{\eta}$ should be equal to one). But in case of analyze of the eigenvectors it's no use and it is allowed to decrease dimension of the vector $\vec{\eta}$.

$$S^* \cdot \vec{\eta}^* = \vec{b}^*, \ \vec{\eta}^* = (A_{11}, A_{22}, A_{12}, B_1, B_2)^T \tag{10}$$

Here constant C is just scale factor of our approximation. Quadratic form A is second rank symmetric tensor. We analyze the collagen orientation using by eigenvalues of the tensor A.

2.3   Algorithm of the Analyze

We used specific algorithm to analyze the micro photos. Generally the algorithm of the photo analyze can be described as:

(1) Determination segment of color vector of the analyzing object
(2) Binarization of the photo
(3) Meshing the photo
(4) For every element we calculate relative content of the object, construct the MIL and it's approximation by ellipse, then find eigenvalues and eigenvectors.
(5) We receive vector field of object distribution.

To identify collagen (paragraph 1 in algorithm) we used equation:

$$||C(i,j) - C_{col}|| < \varepsilon, \tag{11}$$

where C(x, y) – color of bitmap in point (i, j), C_{col} – color of the collagen, ε – color tolerance.

According to color distance photos was binarizied (paragraph 2 in algorithm). It was not classic binarization, because we used three colors: 0 – collagen, 1 – connective tissue, 2 – other not analyzing tissue. Then bitmap was meshed by structured grid and for every element MIL was constructed and approximated by quadratic form (paragraph 4 in algorithm). After that eigenvalues and eigenvectors was calculated and analyzed (paragraph 5 in algorithm). To analyze the quality of the derma we normalized eigenvectors by degree of anisotropy (2). To estimate quality of collagen recovery Q in all area we used equation:

$$Q = \frac{\int\limits_{|\vec{F}(x,y)| \le \delta} dS}{\int\limits_{S} dS}, \tag{12}$$

where $|\vec{F}(x,y)| \le \delta$ mean part of area where magnitude of the vector field is lower than set threshold δ.

3   Results and Discussion

Presented algorithm was used on analyze the quality of derma's recovery. For this purpose we analyzed the micro photos of derma. For analyze in Eq. (11) we used color tolerance equal to 0.05 and in Eq. (12) we used threshold equal to 0.8. On Fig. 1 presented example of original photo (Fig. 1a) and binarized (Fig. 1b). Then the binarized picture was meshed and for every mesh element MIL was build and approximated. On Fig. 2 an example of oriented mesh element presented (white dots - collagen). Red arrow on Fig. 2 shows the first eigenvector (relative to the largest eigenvalue), green arrow shows the second eigenvector (relative to the lowest

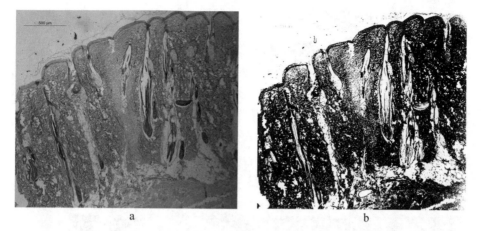

Fig. 1. Example of origin and binarized picture.

eigenvalue). On Fig. 2a presented mesh element with collagen density 2% and degree of the anisotropy 0.27, it means that in the first eigenvector direction elongation is greater by 73% than in the second eigenvector. It shows that orientation of collagen is not chaotic. On Fig. 2b presented mesh element with collagen density 42% and degree of the anisotropy 0.80, it means that in the first eigenvector direction elongation is greater by 20% than in the second eigenvector. This difference is not significant and it means that orientation of collagen is chaotic. Analogic calculations was done for every element of the mesh.

As a result of the calculations vector field can be restored. To simplify analyze, on Fig. 2c vector field was restored for the first eigenvectors normalized by degree of anisotropy. Quality of collagen recovery can be estimated by Eq. (12), for presented results Q was equal to 0.54. It mean that more than half (54%) of collagen objects are oriented (not chaotic) in the presented sample. Of course this parameter has integral nature and it is important to understand areas of chaotic and not chaotic collagens orientation. But for primary quick analyze this parameter is significant. We can specify the area of resulting vector field for detailed analyze. More than, methods for analyze of the result can be easily expanded, because as a result of calculation we got vector field. It means that all mathematical methods to analyze vector fields can be used.

We used our method on samples of three types: no ficin group (N), ficin group (F), immobilized ficin (Fi). We tested 10 bitmaps for every group and we got results for all bitmaps according described technique. Quality of collagen recovery values was: for N group – 48% ± 8%, for F group 78 ± 7%, for Fi group 68 ± 9%. it can be concluded that ficin positively influence on dermas recovery. This effect is known [7, 8] and this is just confirmation of the methodic.

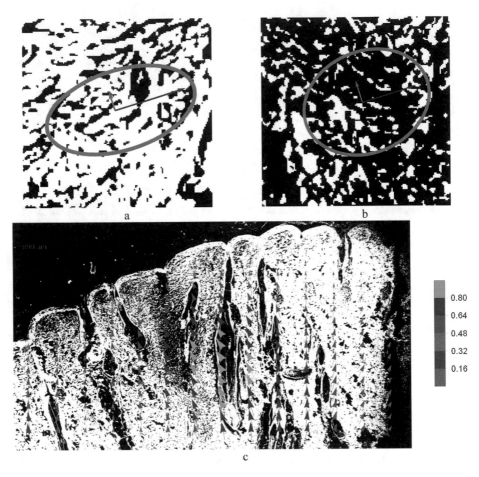

Fig. 2. Illustration of fabric tensor and it's eigenvectors: a – fabric tensor in mesh element with aspect ratio 0.27 and density 90%, b – fabric tensor in mesh element with aspect ratio 0.80, density 68%, c – vector field normalized by aspect ratio with the origin bitmap.

4   Conclusion

Methodic to automatic analyses of microscope snapshots is presented. Object of analysis can be determine by color vector. Then the snapshot can be binarized and meshed. For every element we can restore distribution of the mean intercept length. Orientation of the objects can be calculated using approximation of the mean intercept length. Equation to estimate the quality of collagen recovery was presented. Described technique was used to analyze snapshots of derma. Received results are consistent with published results. This fact shows effectiveness of the method. Traditionally analyze of the collagen distribution is manual and quality of the analyze depends on specialist's experience. Algorithm can be easily transferred in some software for the automatic processing of the picture, for example ImageJ.

Acknowledgements. This work was supported be the grant of the President of the Russian Federation №MK-1717.2018.1

References

1. Yang, L., Liu, Y., Wu, H., Song, Z., Hoiby, N., et al.: Combating biofilms. FEMS Immunol. Med. Microbiol. **65**, 146–147 (2012)
2. Richards, J.J., Melander, C.: Controlling bacterial biofilms. ChemBioChem **10**, 2287–2294 (2009)
3. Donlan, R.M.: Biofilms - microbial life on surfaces. Emerg. Infect. Dis. **8**, 881–890 (2002)
4. Worthington, R.J., Blackledge, M.S., Melander, C.: Small-molecule inhibition of bacterial two-component systems to combat antibiotic resistance and virulence. Future Med. Chem. **5**, 1265–1284 (2013)
5. Blackledge, M.S., Worthington, R.J., Melander, C.: Biologically inspired strategies for combating bacterial biofilms. Curr. Opin. Pharmacol. **13**, 699–706 (2013)
6. Sharafutdinov, I., Shigapova, Z., Baltin, M., Akhmetov, N., Bogachev, M.: HtrA protease from bacillus subtilis suppresses the bacterial fouling of the rat skin injuries. Bionanoscience **6**, 564–567 (2016)
7. Leroy, C., Delbarre-Ladrat, C., Ghillebaert, F., Compere, C., Combes, D.: Effects of commercial enzymes on the adhesion of a marine biofilm-forming bacterium. Biofouling **11**, 11–22 (2008)
8. Baidamshina, D.R., Trizna, E.Y., Holyavka, M.G., Boqachev, M.I., Artyukhov, V.G., et al.: Targeting microbial biofilms using Ficin, a nonspecific plant protease. Sci. Rep. **7**, 46068 (2015)
9. Corr, D.T., Hart, D.A.: Biomechanics of scar tissue and uninjured skin. Adv. Wound Care **2**, 37–43 (2013)
10. van Zuijlen, P.P., Ruurda, J.J., van Veen, H.A., van Marle, J., van Trier, A.J., Groenevelt, F., Kreis, R.W., Middelkoop, E.: Collagen morphology in human skin and scar tissue: no adaptations in response to mechanical loading at joints. Burns **29**(5), 423–431 (2003)
11. Rawlins, J.M., Lam, W.L., Karoo, R.O., Naylor, I.L., Sharpe, D.T.: Quantifying collagen type in mature burn scars: a novel approach using histology and digital image analysis. J. Burn Care Res. **27**(1), 60–65 (2006)
12. Hooman, K., Zhong, Z., Calvin, N., Janette, Z., Xinli, Z., Joyce, W., Kang, T., Chia, S.A.: Quantitative approach to scar analysis. Am. J. Pathol. **178**(2), 621–628 (2011). https://doi.org/10.1016/j.ajpath.2010.10.019
13. Shigapova, F.A., Baltina, T.V., Konoplev, Y.G., Sachenkov, O.A.: Methods for automatic processing and analysis of orthotropic biological structures by microscopy and computed. Int. J. Pharm. Technol. **8**(3), 14953–14964 (2016)
14. Gerasimov, O., Shigapova, F., Konoplev, Y., Sachenkov, O.: Evaluation of the stress-strain state of a one-dimensional heterogeneous porous structure. IOP Conference Series: Materials Science and Engineering, vol. 158, no. 1, p. 012036 (2016)
15. Gerasimov, O., Shigapova, F., Konoplev, Y., Sachenkov, O.: The evolution of the bone in the half-plane under the influence of external pressure. In: IOP Conference Series: Materials Science and Engineering, vol. 158, no. 1, p. 012037 (2016)
16. Harrigan, T.P., Mann, R.W.: Characterization of microstructural anisotropy in orthotropic materials using a second rank tensor. J. Mater. Sci. **19**, 761–767 (1984)
17. Chernov, N., Ma, H.: Computer vision, least squares fitting of quadratic curves and surfaces, pp. 287–302 (2011)

Application of Artificial Neural Networks in the Human Identification Based on Thermal Image of Hands

Tomasz Walczak$^{(\boxtimes)}$, Jakub Krzysztof Grabski ,
Martyna Michałowska , and Dominika Szadkowska

Institute of Applied Mechanics, Faculty of Mechanical Engineering
and Management, Poznan University of Technology,
Piotrowo 3, 60-965 Poznan, Poland
tomasz.walczak@put.poznan.pl

Abstract. The aim of this study was to check the possibility of identifying the persons based on the properties of thermal maps and a temperature distribution of a hand, obtained from a thermal image, with use of artificial neural networks. For this purpose, a series of thermographs of the right hand of eight people was taken, with a thermal imaging camera. The photos were taken under the same thermal conditions, but with different state of warming of hands. After processing the photos (determining the edges, characteristic hand points and areas of interest), the parameters characterizing the metacarpal temperature distribution were determined. Eight parameters were chosen, which were average temperatures of the areas of interest. These parameters were input data of neural networks in the learning and identification process. As it was shown in this study, these parameters were sufficient to clearly identify the persons. Neural networks, designed as multi-layered perceptron, after proper learning showed very high values of identification parameters, including high values of sensitivity and specificity, what proves the high quality of classification. Such identification is possible with the natural thermal state of the hand and if thermal images are not strongly disturbed, the artificial neural networks are very good tool to implement in persons identification process.

Keywords: Human identification · Neural networks · Thermography

1 Introduction

Nowadays recognizing people is a very important problem, because it determines the possibility of identifying and authenticating the subject, which in many everyday activities becomes indispensable. Such issues and many other problems of modern engineering have influenced on the significant development of computational systems. More and more of them use the artificial intelligence algorithms, especially in situations when one deal with issues of a strongly non-linear model or try to analyze data that is incomplete or heavily disturbed. Especially when operating under conditions of incomplete information, the use of artificial neural networks provides greater computing power and better results. Network-based computing systems have a number of

K. Arkusz et al. (Eds.): BIOMECHANICS 2018, AISC 831, pp. 114–122, 2019.
https://doi.org/10.1007/978-3-319-97286-2_11

advantages, which can include automatic learning that do not require knowledge about proper solution of the task, generalization and data generation, and their parallel processing [9, 10, 13]. Identification methods often use artificial intelligence methods, in particular artificial neural networks (ANN). The key issue seems to be finding the right parameters for unambiguous identification of people.

In the tasks of person identification, we can distinguish two types of biometric identifiers: physiological and behavioral. The first ones are based on anatomical features, such as the biometrics based on the structure of the human iris, the shape of the fingerprint, face geometry, hand geometry, or the wrist or finger blood vessel system [6, 7]. Usually they are measured at a certain moment and they are invariant in a long period of time. On the other hand, the biometric identifiers of second type are based on behavioral traits, e.g. biometrics based on hand writing, keyboard typing or gait [4, 5, 10]. They are identifiers learned and acquired over time, and dependent on the current state of mind. They focus on the individual way of doing an activity. However, the physiological characteristics are the most commonly used, due to their effectiveness, versatility and high durability, as well as a relatively small mutability over time.

Human identification based on the temperature distribution of a hand is a new biometric method. It is not widely used for identifying purposes yet, although the results of research indicate the uniqueness of this feature as biometric characteristics. One can find some works confirming the usefulness of thermal imaging in the processes of identifying [7, 11, 12]. However, they mainly concern analyzes based on observation of thermal maps of hands in the aspect of obtaining blood vessel geometry.

Badawi in his work describes the operation of the HVVS verification system (Hand Vein Verification System), which uses the vein system of the upper palm [1]. Recognition process was carried out on the basis of photos obtained with a CCD camera with an infrared filter. Due to the phenomena of infrared (IR) light absorption by hemoglobin, the images were characterized by a high contrast between the blood vessel pattern and the skin tissue, compared to images taken in visible light. The processed images representing the vascular structures were compared with the biometric patterns available in the database. They obtained very good fitting results confirming that the geometry of blood vessels is an individual feature for every man.

Another work devoted to the identification of identities based on thermal images of the blood vessel system is the publication written by Lin and Fan [3]. Researchers used the IR camera to capture images, and they applied the Fourier and Stefan – Boltzmann's laws in the segmentation of the venous patterns. The region of interest (ROI) was determined based on the characteristic points of the hand. The extraction of features was carried out in grayscale, taking different values for the circulatory system and skin tissue. Segmentation was carried out using the watershed segmentation method what allowed to isolate characteristic points of the hand blood system. As a result, the authors obtained a minimum recognition error of 2.3%.

Czajka and Bulwan researched the biometric system based directly on thermal features of hands. Using a specially constructed measurement module, the authors took images of thermal maps, which were then processed to highlight the area carrying biometric information. The object of observation was the distribution of relative temperature of the inner side of the hand [2]. The images were taken using a matrix with thermal sensors, which is a cheaper alternative solution to the traditional thermal imaging cameras. The recognition proceeded on the basis of the thermal maps obtained and included sequentially thermal image segmentation, feature selection and classification. The k-nearest neighbors classifier was used in identification algorithm together with support vector machine. In the work they have gained the equal error rate (EER) 6.67% what proves, that the thermal map distribution of the hands is an individual feature.

However, the method studied in this work is not used as an independent biometric technique, due to limited accuracy. It can be a support for other commonly used methods, such as hand geometry or blood vessel system, and enable the analysis of authenticity based on biometric data.

2 Methods

The main goal of the research was to analyze the properties of thermal maps (hand temperature distribution) in order to identify the human identity. The data was acquired using a thermal imaging camera to check the possibility of creating a biometric system based on thermal images. The effectiveness of biometric identification depends on the features chosen to the description of identity. Occurrence of distortions of a given attribute causes complications in the process of proper recognition. It was necessary to create a measurement system that would enable the collection of the desired features, as well as to determine the procedures of the measurement process that would facilitate their exposure. The primary goal of the measuring station is to achieve the best results in the biometric process. created system consisted of the following parts:

- biometric basis forcing the right positioning of the hand,
- a station with a thermal imaging camera and adapted lighting and ambient temperature,
- a computer with an environment enabling the processing and analysis of the examined features.

Proper collection of the biometric samples was dependent on the precise calibration of the thermal imaging camera and on proper hand positioning. Otherwise, it might affects the display of temperature distributions, which are then used in the processing and analysis. The FLIR T420 thermal imaging camera was used to collect data, allowing thermal images to be taken with a resolution of 240 × 320 pixels. The obtained images were represented in the form of a matrix with raw temperature values, in range from 23.5 °C to 40 °C. The arrangement of the characteristic points of the base has been selected in such a way as to stabilize the hand position during data extraction. Thanks to that, the data acquisition was systematized, what facilitated the acquisition of hand characteristic points, on the basis of designated areas of interests. In addition, the sampling procedure was standardized, by asking each of the person to:

- took a place at the measuring stand in a specific way,
- straighten the wrist and put a relaxed hand on the surface of the base, in such a way as to reveal its inner palm,
- push the thumb and middle finger to the points located on the base.

The database consisted of thermal photographs taken in three series, characterized by different degrees of heating of the metacarpal surface of the right hand. The images were taken for the normal temperature of the hand, hand warmed and cooled. About 90 pictures for each person were taken, and the study group consisted of eight people: 3 men and 5 women. An example of collected thermal map is presented in the Fig. 1.

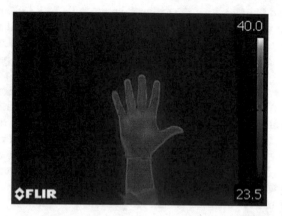

Fig. 1. The example of thermal map of the right hand – normal temperature.

The aim of initial processing of the obtained images was to determine the position of a group of characteristic hand points, as shown in Fig. 2. To do this, the image was first changed into a monochromatic image and then the palm edges were marked using the Sobel filter or the Laplace filter. Sometimes, after filtration the edge fragments were

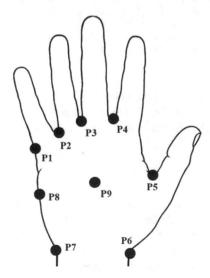

Fig. 2. The characteristic points determined from the hand's edge image.

missing due to the poor quality of the image. These images with missing parts were subsequently removed and were not taken into further consideration.

The eight points possible to determine, regardless of the shape of the hand, were taken under consideration (points from P1 to P8 in the Fig. 2), while point P9 was determined as the center point of the remaining eight. On the basis of those characteristic points, ROI areas were determined, which constituted eight 11 × 11 pixel squares, defined in the middle of the distance between P1 to P8 points, and the center point P9. Due to the fact that the fingers are the part of the hand most sensitive to temperature change depending on external factors, these areas have not been included to the analysis. The analysis was focused exclusively on the middle part of the hand. In further analysis of the metacarpus temperature distribution, for each of the considered eight squares, the ratio of the mean temperature value of each square (average of 121 values) to the maximum value of temperature obtained for the photograph, was calculated. Such normalization allowed the identification, on the basis of the temperature distribution, in considered ROI and not on the basis of only values of temperature field in those areas. It allowed to perform detailed analysis of thermal maps, of individual parts of the hand. Finally, each thermal map obtained after processing was characterized by only eight values from the range <0.1>. The basic statistics of data collected from all ROI for each considered person are presented in Table 1.

Table 1. The mean values representing mean temperature of ROI squares, determined for all 8 person (calculations based on all pictures of single person used in recognition process)

Number of person	Number of ROI							
	1	2	3	4	5	6	7	8
1	0,178 ± 0,06	0,117 ± 0,01	0,100 ± 0,01	0,094 ± 0,01	0,116 ± 0,01	0,535 ± 0,04	0,554 ± 0,04	0,520 ± 0,07
2	0,138 ± 0,10	0,107 ± 0,08	0,087 ± 0,06	0,082 ± 0,05	0,094 ± 0,07	0,305 ± 0,16	0,446 ± 0,15	0,283 ± 0,16
3	0,066 ± 0,01	0,061 ± 0,01	0,065 ± 0,01	0,091 ± 0,01	0,075 ± 0,01	0,066 ± 0,06	0,113 ± 0,09	0,070 ± 0,01
4	0,543 ± 0,06	0,519 ± 0,10	0,464 ± 0,14	0,386 ± 0,15	0,474 ± 0,16	0,545 ± 0,05	0,560 ± 0,04	0,584 ± 0,04
5	0,514 ± 0,11	0,559 ± 0,10	0,597 ± 0,07	0,600 ± 0,06	0,581 ± 0,11	0,595 ± 0,07	0,326 ± 0,11	0,271 ± 0,11
6	0,311 ± 0,03	0,342 ± 0,04	0,392 ± 0,03	0,410 ± 0,04	0,409 ± 0,06	0,210 ± 0,05	0,157 ± 0,03	0,259 ± 0,06
7	0,444 ± 0,04	0,475 ± 0,02	0,496 ± 0,01	0,519 ± 0,02	0,544 ± 0,02	0,382 ± 0,12	0,214 ± 0,01	0,418 ± 0,06
8	0,456 ± 0,11	0,493 ± 0,08	0,520 ± 0,06	0,534 ± 0,07	0,420 ± 0,18	0,316 ± 0,15	0,273 ± 0,13	0,366 ± 0,16

The research was aimed at obtaining data for the non-linear classification process, using the artificial neural networks. Data whose statistics are presented in Table 1 were used as input data for ANN in the training and classification processes.

3   Results

To conduct the classification process on the basis of parameters obtained from thermal maps of hands, a one-way multi-layer perceptron (MLP) was introduced. This is the most commonly used neural network structure, that allows the approximation and generalization of complex mappings, prediction or classification. An important factor affecting the properties of the neural network and the correctness of the learning process is the appropriate network architecture, which consists of the number of perceptron layers and the number of neurons in each of them. Neural networks used in this study consisted of an input layer with 8 neurons and an output layer with one neuron. The number of hidden layers and the number of neurons in hidden layers were variable for different cases. The network's task was to identify people based on the characteristic features of the thermal image of the hand. The properly learned network should response in the form of the value 1 or 0, where 1 indicates the correct identi-fication of the person, while 0 was reserved for the unidentified person. Recognition level of the individuals was assessed by Root Mean Squared Error (RMSE) calculated according to the following formula:

$$RMSE = \sqrt{\frac{\sum_{n=1}^{m}(x - x_n)^2}{m}} \tag{1}$$

where: x – value of the expected response (0 or 1), x_n – network output value, m – number of parameter sets used to test neural network. In addition, to validate created networks as a classifiers, two commonly used metrics were introduced [8]: sensitivity and specificity. These were calculated with the use of two following formulas:

$$sensitivity = \frac{TP}{TP + FN'}$$
$$specificity = \frac{TN}{TN + FP'} \tag{2}$$

where TP means number of true positive classifications (correctly recognized individ-uals), TN is number of true negative classifications (correctly unrecognized individuals), FN is number of false negative classifications (incorrectly recognized individuals) and FP is a number of false positive classifications (incorrectly unrecognized individuals). The bigger is value of sensitivity and specificity the classifier is better one.

As it was mentioned in Sect. 2, at the input of each network, the 8 parameters of normalized average temperatures of the designated eight areas of interest were given. The output was one neuron with a linear activation function and in the hidden layer there were from 2 to 4 neurons. Because of the use of linear function in output (not binary step function, like in typical classification approach), the final response of identification system was calculated according to the following principle:

- when the network response was within the range $(-\infty, 0.5)$, the result was rounded to 0 (person unrecognized),
- when the network response was in the range $<0.5, \infty)$, the result was rounded to 1 (person identified).

All simulations were conducted with Matlab environment together with Neural Net-work Toolbox.

For further analysis, the network which achieved the best results on the training set was selected for each person. Training of networks were conducted with Levenberg-Marquardt method with data split into three subsets: 70% - training set, 15% - vali-dation set, 15% - test set. The results of learning processes were measured with $RMSE$,

Table 2. The recognition results (*RMSE*), number of wrong answers (*FP + FN*), *sensitivity* and *specificity* of classifiers for each considered person.

Number of person	RMSE	FP + FN	Sensitivity [%]	Specificity [%]
1	0,0179	4	100,00	98,10
2	0,0352	8	76,67	99,52
3	0,0414	8	83,33	98,57
4	0,0363	9	96,67	96,19
5	0,0042	1	96,67	100,00
6	0,0253	19	83,33	93,33
7	0,0374	8	93,33	97,14
8	0,0407	10	73,33	99,00

and in all chosen networks to further considerations the value of errors were less than 0.08 for considered three subsets. The results of neural networks recognition for each of the eight examined person are summarized in Table 2.

All results that are presented in Table 2 were obtained by processing all 240 thermal pictures taken in normal condition of a hand. As one can observed all designed neural networks are seemed to be good quality classifiers. The worst one generated 19 wrong answers but it still has high specificity and sensitivity, and should process data in a proper way.

4   Discussion and Conclusions

The quality of radiation emitted by a body is influenced by many factors. It depends not only on the temperature of the tested object, but also on the state of its surface. The thermal imaging camera perceives the radiation of a examined body, but also records the emission from its surroundings or reflected rays. The atmosphere, and in particular the water vapor or carbon dioxide contained in it, suppress the radiation causing a decrease in the accuracy of the measured temperature distribution of the tested object. To minimize measurement errors, attention should be paid to such factors as the emissivity of the subject being analyzed, the ambient temperature, the relative humidity of the atmosphere and the distance between the tested object and the camera. When measuring temperature, reduce the impact of unwanted radiation sources and determine the appropriate parameters of the analyzed entity. This will enable automatic adjustment of the thermal imaging camera to the existing operating conditions and compensation of the influence of disturbances.

Modern thermal imaging cameras are characterized by better accuracy, sensitivity and resolution. They enable temperature measurement and visualization of its distribution in the form of thermal maps in a wide range of colors. They give a chance to determine the isotherms and individual measurement points and to store many images. Biometric systems using thermal images provide a high level of security with simultaneous convenience. The measuring method does not require direct contact of the user with the device, it is a hygienic and non-invasive solution, which is why it is distinguished by a high level of user acceptance. The main advantage of thermograms is their independence from lighting conditions [11].

Studies carried out in the work on thermal images of hands indicate that the temperature distribution is an individual feature of each human being. Its biggest disadvantage is the ability to easily modify the thermogram of a person which is being recognized. One can do it by even rubbing hand or holding a cold or hot object in the hand. Tests were also carried out on excessively warmed and cooled hands. Classification results even for much more complex architectures of neural networks were far from satisfactory (with sensitivity and specificity below 50%). Therefore in this work only results obtained for natural state of hand were presented.

The recognition efficiency seems to be very good, but to clearly formulate thesis about possibility of application of this method, much bigger group of person should be taken into consideration in further research. The results presented in Table 2 shows, that all eight persons were identified by neural network with small RMSE value, and high specificity of each classifier. However, it can be observed that for people with the highest temperature variability in ROI areas (see Table 1, person number 2 and 8), the sensitivity values are relatively low. For some person, the temperature change on the hand surface is extremely fast, e.g. due to the movement of the fingers or holding something for an instant. This usually leads to a much more distorted thermal image. Therefore, the necessary condition for the correct identification is the most "natural state" of the hands when taking measurements. While maintaining this condition, hand thermography can be used as an independent biometric method or a method that supports other known techniques.

Acknowledgment. The work was funded by the grant 02/21/DSPB/3493 from the Ministry of Higher Education and Science, Poland.

During the realization of this work Dr. Jakub K. Grabski was supported with scholarship funded by the Foundation for Polish Science (FNP).

References

1. Badawi, A.: Hand vein biometric verification prototype: a testing performance and patterns similarity. In: Proceedings of the 2006 International Conference on Image Processing, Computer Vision, and Pattern Recognition, IPCV 2006, Las Vegas, Nevada, USA (2006)
2. Czajka, A., Bulwan, P.: Biometrics verification based on hand thermal images. In: Proceedings of the 6th International Conference on Biometrics, ICB 2013, Madrid, Spain (2013)
3. Fan, K., Lin, C.: The using of thermal images of palm-dorsa vein-patterns for biometric verification. IEEE Trans. Circ. Syst. Video Technol. **14**(2), 199–213 (2004)
4. Grabski, J.K., Walczak, T., Michałowska, M., Cieślak, M.: Gender recognition using artificial neural networks and data coming from force plates, innovations in biomedical engineering. In: Gzik, M. et al. (eds.) IBE 2017. Advances in Intelligent Systems and Computing, vol. 623, pp. 53–60. Springer, Cham (2018)
5. Jain, A.K., Bolle, R., Pankanti, S.: Biometrics: Personal Identification in Networked Society. Kluwer, Norwell (2006)
6. Jain, A.K., Flynn, P., Ross, A.A.: Handbook of Biometrics. Springer, New York (2008)

7. Kumar, A., Hanmandlu, M., Madasu, V.K., Lovell, B.C.: Biometric authentication based on infrared thermal hand vein patterns. Digit. Image Comput. Tech. Appl. **2008**, 331–338 (2009)
8. Fawcett, T.: An introduction to ROC analysis. Pattern Recogn. Lett. **27**, 861–874 (2006)
9. Peng, H., Long, F., Ding, C.: Feature selection based on mutual information criteria of max-dependency, max-relevance, and min-redundancy. IEEE Trans. Pattern Anal. Mach. Intell. Pattern Anal. Mach. Intell. **27**(8), 1226–1238 (2005)
10. Walczak, T., Grabski, J.K., Grajewska, M., Michałowska, M.: The recognition of human by the dynamic determinants of the gait with use of ANN. In: Awrejcewicz, J. (ed.) Springer Proceedings in Mathematics and Statistics, Dynamical Systems: Modelling, vol. 181, pp. 375–385. Springer, Cham (2016)
11. Wang, L., Leedham, G.: A thermal hand vein pattern verification system. In: Singh, S., Singh, M., Apte, C., Perner, P. (eds.) Pattern Recognition and Image Analysis. Lecture Notes in Computer Science, vol. 3687, pp. 58–65. Springer, Berlin, Heidelberg (2005)
12. Wang, M.H.: Hand recognition using thermal image and extension neural network. Math. Prob. Eng. **2012**, 15 (2012)
13. Xiao, Q.: A note on computational intelligence methods in biometrics. Int. J. Biometr. **4**(2), 180–188 (2012)

Impact/Injury Biomechanics

Stability of the EMG Signal Level Within a Six-Day Measuring Cycle

Robert Barański$^{(\boxtimes)}$ (iD)

AGH University of Science and Technology,
al. Mickiewicza 30, 30-059 Kraków, Poland
robertb@agh.edu.pl

Abstract. This paper presents the results of research on the identification of changes in the electromyographic (EMG) signal recorded with the surface method (sEMG) over the course of six consecutive days. The signal was recorded for two people. The electrodes were fixed on the upper limb in two places of the forearm (over the brachioradialis and the bully of the superficial flexor (flexor digitorum superficialis). Muscles were activated by the hand clamp on the handle in the range of $25 \div 100$ N. 21 measurement series were analysed, which consisted of 966 individual clamps. Estimates like root mean square, average value, energy and turn per second were used for the research. Due to the lack of a normal distribution of the estimators, non-parametric tests were performed in most cases. The tests carried out did not allow us to infer about the lack of changes in the signal over the period of six days under investigation. Moreover, by shortening the period of the tested series even to several successive series of measurements, it was also impossible to determine consistent conclusions for all the tested forces. Registered signals were characterized by very high variability between particular series. What's more, the correlation studies between changes of individual forces per day also do not support the hypothesis that there is a constant, time-independent measurement of the relationship between the recorded EMG signal and the force.

Keywords: EMG · sEMG · Force estimation · Signal analysis
EMG measurement

1 Introduction

Study of dependencies that combine muscle work and the electromyographic signal (EMG) generated by them is intensively conducted since the 1960s [1]. In many scientific centres different approaches to this issue have been used. Part of the work focuses on the usefulness of EMG signals in early (screening) medical diagnosis [2], others on the rehabilitation process after injuries [3,4], still others on the dependence of the intensity of the EMG signal on the condition of the muscle. Among the latter, one can distinguish between works focusing on the

© Springer Nature Switzerland AG 2019
K. Arkusz et al. (Eds.): BIOMECHANICS 2018, AISC 831, pp. 125–137, 2019.
https://doi.org/10.1007/978-3-319-97286-2_12

assessment of the force realized by the muscle [5–7] and work focusing on the assessment of the degree of muscle fatigue during the implementation of the tasks [8–10].

This work is related to a wider project aimed at estimating clamping force based on the EMG signal. The reason for taking up this topic is an attempt to provide a tool to meet the requirements of the European standard [11], that imposes the need to measure hand contact forces with the tool when determining human exposure to vibrations transmitted by the upper limbs. Currently, this requirement is neglected due to large measurement problems, which would not influence the operation of the tool as well as the transmission of vibrations from the tool to the operator's body.

Anatomical analysis of the muscles involved in the operation of hand clamping on the tool handle allows to conclude that it is an activity characterized primarily by the involvement of the frontal forearm musculature. The muscles directly involved in this process include the flexor digitorum superficialis muscle (FDS), being the strongest flexor of the hand responsible for bending the fingers; the multi-fingered flexor muscle (FDP) supporting FDS; radial flexor muscles (FCRs) and elbow wrists (FCUs), which stabilize the wrist while pinching the fingers [12].

In many scientific centres, work is being carried out to determine the functional dependence between muscle force and tension, but until now no unified universal transition function has been developed for this problem. The main reason is individual factors of the subjects which, being unique for each person, make it impossible to develop a universal transition function for EMG - force dependence. The question then arises whether it is possible to develop a personalized EMG - force transfer function. Does it change over time?

Among the works related to longterm EMG signal research, health publications are the most popular ones. Some of them are focused on the study of changes in the signal depending on the degree of disease status [13,14]. Others focus on rehabilitation [15]. One of the work combining the element of influence of the interval of measurement and the force of the muscle tension is [16]. In this work, measurements recorded at intervals of 3 min, 90 min and 6 weeks were used to analyse the EMG leg signal. However, these were not measurements in which the point of interest was the dependence of the EMG signal - force (expressed in Newtons), based mainly on the parameter maximum voluntary contraction (MVC). Although the work proves the influence of time on the stability of the analysed signals, it does not answer questions whether the changes occur in a predictable way.

There were no works that would regularly examine the behaviour of the EMG signal in correlation with force in hours or days. Therefore, this work tries to fill this gap. The focus was on examining the changes in the EMG signal recorded over the next six days. In most cases, two series of measurements were made daily.

2   Methods

Measurements and analysis of estimators were made using the LabVIEW 2016 environment. Statistical analyses were performed in the Statistica 13.1 environment.

2.1   Measurement Procedures

The tests were carried out in six consecutive days. In order to ensure repeatability all measurements were carried out on the same station (same laboratory). The test was carried out for a sitting position. The upper limb was supported in the forearm, bent at the elbow joint at an angle of about 120°. This setting allows you to eliminate the forearm reversing-reversal function created with the participation of FCR and FCU. The tests were carried out for two healthy people. Subjects provided written consent to do the measurements. The first person (designated as I) is a man, 41 years old, 185 cm tall, weight 84 kg. The second person (designated as II) is a woman, age 63, 169 cm tall, weight 79 kg. The decision to differentiate the characteristics of examined persons enabled checking the influence of this parameters on results (e.g. person I has a strong repeatability of EMG signals with a complete lack of repeatability for a person II).

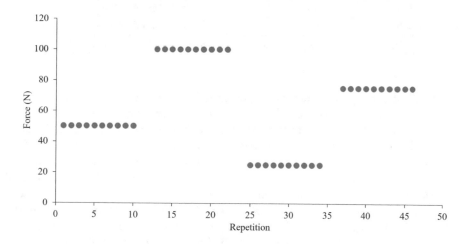

Fig. 1. Measurement sequence for one series

Measurements were carried out in series. Each series is a sequence of four measurements for previously defined forces. The measurement for each force consisted of a sequence of minimum 10 clamps with a fixed set force lasting 6 s and a rest not shorter than 10 s. In the case of the man, the forces were 25, 50, 75 and 100 N. For the woman the range of forces was reduced to 25 N, 40 N, 50 N and 75 N. The order of forces during the measurement series was random (e.g. 50 N, 100 N, 25 N, 75 N) to avoid the situation of getting the muscle accustomed to the increasing level of force (excitation). The interval between the series of measurements was several minutes. An exemplary measurement sequence is shown in Fig. 1.

During the measurements the tested person constantly monitored the level of force, and her task was to maintain the force with an accuracy of $\pm 5\%$ of the set point. Only the repetitions that meet this condition were taken into account. 11 measuring series for a man and 10 series for a woman were made, which in total made up to 966 repetitions of hand clamps on the handle. The Table 1 below presents the occurrence of series on particular days. In the case of two series a day, they were carried out for at least 6-h. The empty fields mean no data for analysis in the selected measurement series.

Table 1. Study plan, series and force distribution (a - 25 N, b - 40 N, c - 50 N, d - 75 N, e - 100 N)

Day no.	1	2		3		4		5		6	
Series no.	1	2	3	4	5	6	7	8	9	10	11
Person	Electrode position										
I A					a de	acde	acde	acde	acde	acde	acde
B	acde		acde	acde	acde	a de	acde	acde	acde	acde	acde
II A	bc			abcd	abcd	abcd	abcd	abcd	abcd	a cd	
B	bc		abcd	abcd	abcd	abcd	abcd	abcd	abcd	a cd	

The electromyographic signal was obtained by surface method (sEMG). For this purpose, pre-gel electrodes (Ag/AgCl) with a distance of 2 cm between the centres of conductive agents were used. The electrodes were fixed on the forearm in two places. The reference electrode was attached to the elbow joint. The first place (electrode A) was determined on the basis of previous studies [17] and located above the belly brachioradialis at a distance of 1/3 of the interval between the epicondylus medialis humeri and elbow ulcer (processus styloideus ulnae). The second place the electrodes were attached to (electrode B) was the flexor digitorum superficialis. The deep-finger flexor has its origin in 2/3 proximal frontal area and the medial sternum of the elbow and the intercostal membrane. Electrode attachment points were fixed at 1/3 of the distal forearm between the long musculus palmaris longus and the musculus flexor carpi ulnaris. It should be noted that the designation of the said tendons in the subject was carried out in a palpitating manner at the wrist of the palm at about 30° (hand palm length) and flexion of the wrist elbow about 30° (muscle elbow wrist flexor) [18].

The handle has an external dimension of 30 mm × 60 mm (shorter rounded side with a radius of 15 mm) Fig. 2.

The EMG signal conditioning system is a proprietary construction based on the use of two stage amplification (INA128P circuit, CMRR coefficient = 120 dB)

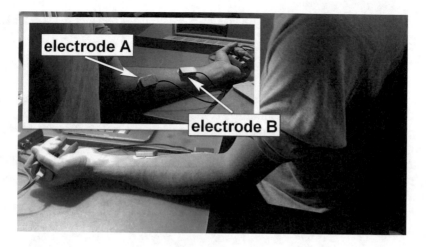

Fig. 2. Upper limb and electrode position during measurements)

and active filtration in the range of 6 Hz – 3000 Hz. A gain of 1200 times was applied. The data was recorded using a NI DAQ6212 multichannel card (16 bit) with a sampling rate of 10 kHz enabling simultaneous measurement of all recorded signals. The software for recording, managing signals and performing analysis was developed using the LabVIEW 2016 environment.

2.2 Methods

In order to perform statistical analysis, it was necessary to calculate an estimator for a time signal that would make it possible to determine the EMG signal level in one value during the clamping time. Four estimators were selected for the analysis. Three of them represented methods of analysis based on signal energy. Some of them were used in the works [13]. The parameter that can be included in the frequency analysis was also used (turn per second). The EMG signal was first subjected to band-pass filtration in the frequency range 10–500 Hz suggested by the ISEK standards [19]. The formulas for calculating individual estimators are presented below. Each time, the variables correspond to respectively: n – the number of samples of the input time series; x_i - the value of the i-th measurement of the EMG signal. The root mean square rms value was calculated as the energy estimator, according to the relationship:

$$rms = \frac{\sum_{i=1}^{n} x_i}{n}.$$

(1)

The next estimator was the mean value after rectification $mean(abs)$ calculated according to the dependence:

$$mean(abs) = \sqrt{\frac{\sum_{i=1}^{n} x_i^2}{n}}.$$

(2)

The last element from the group of energy estimators was energy E calculated according to relationship:

$$E = \sum_{i=1}^{n} [log(x_i^2)].\tag{3}$$

Among the methods based on frequency analysis a turn per second TPS estimator was used. This estimator is calculated based on an algorithm that first differentiates the time signal and then analyses it point by point checking if the signal has passed the zero value.

The selection of estimators E and TPS was dictated by the unpublished results of the research on determining the most stable estimate and function to describe the relationship between the clamping force and the EMG signal. The rms and $mean(abs)$ estimators were chosen because of their versatility and their use in many scientific works.

3   Results

Using the above presented estimators, statistical tests were provided to investigate whether registered EMG signals can be considered stable over the entire series of measurements. The analyses were carried out with the division into persons, the electrodes place and value of the clamping force. All tests were done in a similar way. Each time during the statistical tests the confidence level (p-value) of 0.05 was assumed as a critical value.

3.1   Clamping Force Stability

Due to the fact that the stimulus to generate the EMG signal was a hand clamp on the handle, its significant fluctuations could have influenced the results of further analyses. Therefore, the level of changes in the registered force in each of the measurement series was checked first. The subject task was to maintain a constant clamping force during the measurement series. The permissible force change could not exceed $\pm 5\%$ of the set value. It was checked how the clamping force varied during particular series (the series consisted of 10 to 15 clamps with given force). For this purpose, the percentage coefficient of variation of the clamping force for each series was calculated. In the case of a person I, it took values in the range of $0.13\% \div 0.61\%$. In the case of a person II, it took values in the range of $0.22\% \div 1.29\%$.

Therefore, it can be concluded that the force was maintained steadily during measurements and its changes should not have a significant impact on the recorded EMG signals and thus on the results of the analyses.

3.2   Series Stability

The main element of the research was to determine the significance of the differences in the sets of calculated estimates between series (for each of the given forces independently). The Fig. 3 presents an example chart of distribution changes obtained for the *rms* estimator for person I. Analogous distributions were determined for other estimators and forces.

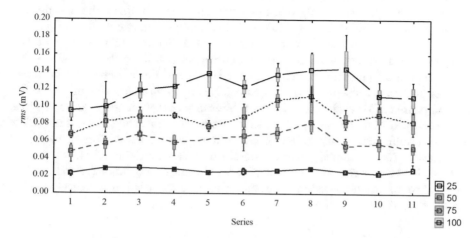

Fig. 3. Changes of the *rms* estimator for all series for person I (point - average, box - standard deviation, mustache - min, max)

In order to perform tests on the statistical significance of population differences between the series of measurements (ANOVA), tests for conformity of variable distributions in the series with the normal distribution were carried out in the first place for each series. As a result, out of 556 individual series (each of them consisted of a dozen hand clamps) in 63 cases there was no basis for recognizing that the distribution of estimators is normal. As a result, out of the tested 64 sets of series only in 28 sets all series there were no grounds to reject the hypothesis of compliance with the normal distribution. It should be noted that the set of series formed series for each person, clamping force, electrode position and the estimator independently. Thus, only in these 28 cases the next step was justified, that is, the homogeneity of the variance. For this purpose the Brown-Forsyth test was used. Only for the 22 sets of the series there were no grounds to reject the null hypothesis about the difference between the variances of the tested series. For them, a parametric one way ANOVA analysis was performed. It turned out that only for the estimator E (energy) for person I, electrode position A and clamping force 100 N there was no basis for rejecting the hypothesis of statistical non-significance of differences between series.

Due to the failure of not meeting the condition regarding the population distribution of the estimators, non-parametric analysis was performed. For this purpose, the Friedman test was used, being a non-parametric counterpart of a one-way analysis of variance for repeated measurements. In the Friedman test, the zero hypothesis assumes that the data contains samples taken from the same population (with equal medians) [20]. The analyses were carried out for all measurements combinations. Also in this case, similar results were obtained. Only for person I, electrodes position A and clamping force 100 N, the lack of differences between sets was statistically significant. However, the difference was to obtain such a result for three estimators: rms, $mean(abs)$ and E.

Due to the fact that only one measurement series fulfilled the assumption made at the beginning, it was also decided to analyse shorter strings of a series of signals (up to now, these were sequences of maximum 11 series). Post hoc tests were carried out, allowing for the analysis of the significance of statistical differences of all possible combinations for the tested measurement series (even for the next two consecutive measurement series). The focus was on the series that appeared after each other. Lack of differences between the examined series was confirmed in many cases, however, it was so rare that the tests did not allow for the formulation of any coherent dependence for all tested forces, electrode positions or the examined person. Therefore, it can be assumed that the time interval between series at the level of a few hours is too large for some event which is to maintain consistency between the data recorded and future-proofing based on them.

3.3   Correlations

This section presents the results of analyses using correlation techniques. As will be further demonstrated, most were based on Spearman's non-parametric rank correlations.

Correlations Between Forces

It was examined if there is a correlation between series sets determined for different forces (the mean values of EMG estimators were compared). Assuming that there is a constant (over several days) relationship between the EMG signal and the force of the clamp, this dependence should be captured using the correlation analysis. Changes over several days should take place proportionally and affect each of the forces in the same way. Significantly, if the discussed dependence of the EMG signal and the clamping force is not constant over time and changes proportionally (regardless of whether it will be linear proportionality or not), this phenomenon will be captured.

It should be emphasized that the phenomenon described in the previous item, i.e. the lack of grounds for inferring the lack of difference in statistical significance between series does not contradict the phenomenon of correlation described above. It is a completely different effect, which needed to be investigated.

Based on the previous tests (no normality of the distributions of estimators in particular series), it was decided to perform the analysis of non-parametric Spearman's ranks correlation. The use of Spearman's ranks had one more advantage because, due to its nature, it also allows nonlinear relationships to be studied, and it is precisely this type of suspicion that the relationship between clamping force and EMG signal is suspected. The obtained results are presented in the Table 2.

Values indicate the strength of correlation, but more importantly, only those marked in red are statistically significant (they should be taken into account). The results obtained allow for making several conclusions.

In a few cases, the statistical significance of the calculated correlation coefficient can be talked about, but there is a tendency that for measurements with the use of electrodes fixed in place B, a larger number can be noted (for both person I and II). In addition, this significance is less sensitive to a change in the estimator (confirmation can be the four cases for which the result is significant regardless of the type of estimator).

Table 2. Spearman's correlation between forces

el.position		A				B			
Person	Forces	rms	$mean(abs)$	TPS	E	rms	$mean(abs)$	TPS	E
I	25–50	0.09	0.09	0.54	0.31	0.44	0.54	0.01	0.56
	25–75	0.68	0.68	0.71	0.68	0.24	0.4	0.42	0.37
	25–100	0.29	0.29	0.71	0.29	0.01	0.19	0.33	0.17
	50–75	0.49	0.49	0.94	0.68	0.87	0.87	0.84	0.82
	50–100	0.26	0.26	0.94	0.29	0.59	0.6	0.93	0.58
	75–100	−0.07	−0.07	1	−0.07	0.45	0.51	0.9	0.49
II	25–40	0.74	0.74	−0.2	0.74	0.81	0.81	0.77	0.88
	25–50	0.54	0.57	0.74	0.57	0.38	0.38	0.33	0.38
	25–75	0.86	0.86	0.61	0.86	0.78	0.77	0.63	0.77
	40–50	0.24	0.24	0.61	0.24	0.5	0.47	0.07	0.43
	40–75	0.26	0.26	−0.03	0.26	0.74	0.67	0.4	0.69
	50–75	0.61	0.74	0.36	0.74	0.85	0.87	0.8	0.87

Despite the lack of statistical significance of correlation coefficients, the calculation of the mean correlation value with regard to the point of electrodes position and tested person was calculated. For both subjects, for the electrodes position B, the average correlation value was higher by about 0.1. However, it should be remembered that this result can only be interpreted as a trend due to the lack of statistical significance.

Person to Person Correlations

It was also checked whether there is a relationship between changes in estimators for both tested persons. The existence of such a relationship could provide, for example, the influence of the measurement procedure or other uncontrolled factors that both person were subjected to. Also, in this case the non-parametric correlation analysis of Spearman's rank was used. Correlations were checked independently for each of the forces 25 N, 50 N and 75 N. Based on the results of the above-presented tests and the statement, and that there are no statistically significant correlations between the results obtained for individual forces over the period of a few days, it can be assumed that in this case there should be no dependencies, too.

The analysis carried out confirmed this assumption. The obtained correlation coefficients did not exceed the value of 0.5, while being burdened with the lack of statistical significance.

3.4   Time Influence of EMG-Force Transfer Function

Determining the function that will describe the dependence of changes between the EMG signal and the clamping force $F = f(EMG)$ is one of the main goals for scientists studying EMG signals. Therefore, in addition to determining the function itself, it is equally important to know what changes in a given function (function coefficients) may be subject to.

In this section, it has been checked how the estimators of the function describing the curve of dependence between the force generated by the EMG signal change over the measurement series. Just like at work [21] analyses were performed for the 2nd order polynomial expressed in dependence (4):

$$f = ax^2 + bx + c. \tag{4}$$

where:
a,b,c - coefficients.

In this way, the coefficients of polynomial a, b, c for the persons studied and the electrodes positions were determined for the rms estimator. Sample results of changes are presented in Fig. 4.

All determined coefficients were characterized by significant and irregular nature of changes. This was confirmed by very high values of the percentage coefficient of variation (V_x), reaching on average a value of several hundred percent.

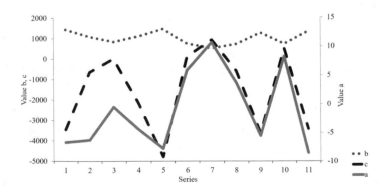

Fig. 4. Changes of coefficients of the 2nd order polynomial (for person I, rms estimator, electrode position B)

In Fig. 4, a very strong correlation can be observed between changes in parameter a and c. The Pearson correlation coefficient for the parameters a and c was determined with the division into persons and places of fixing the electrodes. As expected, it was very high at 0.79 to 0.97 (all statistically significant). This may indicate that the shape pattern of the second degree polynomial describing the dependence of EMG-Force is maintained. However, further research is necessary to better understand this phenomenon (which is not the subject of this work).

4 Discussion

This paper presents the results of research focusing on determining the nature of EMG signal changes over several days. For this purpose, two people were subjected to tests carried out for 6 consecutive days (most of the measurements were carried out twice a day with an interval of at least 6 h). During the tests the EMG signal was measured in two places on the forearm. Measurements were made during a static hand clamp force on the handle (for four previously chosen forces).

The analyses did not give right to the development of a function that would de-scribe the trend of EMG signal changes over time. The non-parametric Friedman test used to investigate the statistical significance of the differences of many dependencies, in one case alone, did not justify rejecting the hypothesis that the recorded signals come from the same population. This is evidenced by the very high variability of the EMG signal over days and even hours.

Another issue examined was the knowledge on the stability of EMG signal proportions for different clamping force values. It was assumed that even in the case of strong uncontrolled external factors affecting the EMG signal recorded, the fluctuations will concern the entire tested range of forces. In other words, the changes will have a proportional effect on the recorded EMG signal values for all tested forces. Such behaviour should result in a high similarity of changes in signals registered at similar times for different forces. To this end, cross correlation coefficients were calculated between the four examined forces. The results obtained were only partially covered by the assumptions. For some cases, very high correlation has been demonstrated (>0.7). However, when analysing the results, it can be seen that it was different for the persons examined and occurred in one in six (person I, rms) and four in six (person II, rms) of the examined couples. The correlation result was strongly dependent on the estimator, an example of which is a significant difference in the obtained results for the rms and TPS estimators.

The last examined element strongly associated with the stability of EMG signal proportionality discussed in relation to the clamping force is the change of the parameters of the function describing the dependence of the EMG signal and the clamping force. This stability was checked for the function of the 2nd order polynomial, by determining the coefficients of variation for individual polynomial coefficients. Values of the determined V_x coefficients reached several hundred percent for all tested persons and the place of attaching the sensors. This result coincides with the observations from the previous paragraph.

It is noted, however, an important feature as manifested by very high (>0.79), the correlation between the parameters a and c polynomial. It has performed for all cases studied.

Funding. This work was supported by the AGH University of Science and Technology [grant number 11.11.130.734].

References

1. Basmajian, J.V.: Muscle Alive: Their Functions Revealed by Electromyography. Williams & Wilkins, Baltimore (1963)
2. Bohannon, R.W.: Hand-grip dynamometry predicts future outcomes in aging adults. J. Geriatr. Phys. Ther. **31**, 3–10 (2008)
3. Shechtman, O., Gestewitz, L., Kimble, C.: Reliability and validity of the DynEx dynamometer. J. Hand Ther. **18**, 339–347 (2005)
4. Lagerström, C., Nordgren, B., Olerud, C.: Evaluation of grip strength measurements after Colles' fracture: a methodological study. Scand. J. Rehabil. Med. **31**, 49–54 (1999)
5. Duque, J., Masset, D., Malchaire, J.: Evaluation of handgrip force from EMG measurements. Appl. Ergon. **26**, 61–66 (1995)
6. Hoozemans, M.J.M., Van Dieën, J.H.: Prediction of handgrip forces using surface EMG of forearm muscles. J. Electromyogr. Kinesiol. **15**, 358–366 (2005)
7. Oskouei, A.H., Carman, A.: Prediction of hand grip force using forearm surface displacement. J. Biomech. **45**, S513 (2012)
8. Roman-Liu, D., Bartuzi, P.: The influence of wrist posture on the time and frequency EMG signal measures of forearm muscles. Gait Posture **37**, 340–344 (2013)
9. Barandun, M., von Tscharner, V., Meuli-Simmen, C., Bowen, V., Valderrabano, V.: Frequency and conduction velocity analysis of the abductor pollicis brevis muscle during early fatigue. J. Electromyogr. Kinesiol. **19**, 65–74 (2009)
10. Roman-Liu, D.: The influence of confounding factors on the relationship between muscle contraction level and MF and MPF values of EMG signal: a review. Int. J. Occup. Saf. Ergon. **22**, 77–91 (2016)
11. International Standard Organisation, ISO 5349-1:2001: Mechanical vibration – measurement and evaluation of human exposure to hand-transmitted vibration – Part 1: general requirements (2001)
12. Reicher, M., Bohenek, A. Anatomia ogólna: kości, stawy i wiezadła, mieśnie (Human anatomy: bones, joints and ligaments, muscles). Wydawnictwo Lekarskie PZWL (2016)
13. Gruet, M., Vallier, J.M., Mely, L., Brisswalter, J.: Long term reliability of EMG measurements in adults with cystic fibrosis. J. Electromyogr. Kinesiol. **20**, 305–312 (2010)
14. Breit, S., Spieker, S., Schulz, J.B., Gasser, T.: Long-term EMG recordings differentiate between Parkinsonian and essential tremor. J. Neurol. **255**, 103–111 (2008)
15. Pylatiuk, C., et al.: Comparison of surface EMG monitoring electrodes for long-term use in rehabilitation device control. In: IEEE International Conference on Rehabilitation Robotics, ICORR 2009, pp. 300–304. IEEE (2009). https://doi.org/10.1109/ICORR.2009.5209576
16. Kollmitzer, J., Ebenbichler, G.R., Kopf, A.: Reliability of surface electromyographic measurements. Clin. Neurophysiol. **110**, 725–734 (1999)

17. Barański, R., Kozupa, A.: Hand grip-EMG muscle response. Acta Phys. Pol. A **125**, A-7–A-10 (2014)
18. Netter, F.H., Thompson, J.C., Dziak, A., Kamiński, B.: Atlas anatomii ortopedycznej Nettera (Netter's Concise Atlas of Orthopaedic Anatomy). Elsevier Urban & Partner (2007)
19. Morletti, R.: Standards for Reporting EMG Data. J. Electromyogr. Kinesiol. **39**, I–II (2018). Politecnico di Torino, Italty
20. Stanisz, A.: Przystepny kurs statystyki z zastosowaniem STATISTICA PL na przykładach z medycyny (Statistic course using STATISTICA software on examples of medicine). StatSoft Polska (2006)
21. Doheny, E.P., Lowery, M.M., FitzPatrick, D.P., O'Malley, M.J.: Effect of elbow joint angle on force-EMG relationships in human elbow flexor and extensor muscles. J. Electromyogr. Kinesiol. **18**, 760–770 (2008)

Prediction of the Segmental Pelvic Ring Fractures Under Impact Loadings During Car Crash

Tomasz Klekiel[1]($\boxtimes$), Katarzyna Arkusz[1], Grzegorz Sławiński[2], and Romuald Będziński[1]

[1] Biomedical Engineering Division, University of Zielona Gora, Licealna 9 Street, 65-547 Zielona Gora, Poland
t.klekiel@ibem.uz.zgora.pl
[2] Faculty of Mechanical Engineering, Department of Mechanics and Applied Computer Science, Military University of Technology, Gen. Witolda Urbanowicza 2 Street, 00-908 Warsaw, Poland
https://www.zib.wm.uz.zgora.pl

Abstract. The Pelvis is the most susceptible part of the body to damage during car accidents and is characterized by the highest mortality rate, especially in the case of multiple fractures. The mechanism of these fractures remains unclear and this makes the development of effective crash protection more difficult. A geometric model of the lumbo-pelvic-hip complex (LPHC) including elements of skeletal, muscular and ligament structure stabilizing the pelvis was elaborated on the computed tomography images of a 25-year-old patient. The influence of pelvic boundary conditions on the type of injuries was subjected to analysis using the Finite Element Method (FEM). The cases of a model anchorage dependent on the position of the passenger's body in the vehicle, where the impact of interior vehicle elements, such as seat belts or a car seat, were taken into account. A fracture threshold was established by applying lateral loads from 0 to 10 kN to a greater trochanter of the femoral bone in each of a five cases of boundary conditions reflecting the influence of different car parts on a passenger's body. The magnitude of the contact force between the body and the vehicle parts during a side collision against the driver's door were determined using the elaborated model. Furthermore, a pelvis lateral collision theory model was built and validated with the use of clinical data. The obtained results can provide an estimate for a threshold of the initial failure in the pelvis bone due to an impact compression transmitted through an overlying tissue. Therefore, it was assumed that the properties of the fractured structure are similar to the cancellous bone.

Keywords: Pelvic injury · Finite elements · Injury mechanism
Fracture of bone · Soft tissue · Car accident

© Springer Nature Switzerland AG 2019
K. Arkusz et al. (Eds.): BIOMECHANICS 2018, AISC 831, pp. 138–149, 2019.
https://doi.org/10.1007/978-3-319-97286-2_13

1   Introduction

Car accidents claimed nearly 3500 lives each day in 2012 made among the ten leading causes of injuries [1]. It is estimated that the injuries resulting from road traffic accidents will be the third most common cause of human deaths in 2030 - despite the best attempts of road safety research. Among the accident injuries, the lumbo-pelvic-hip complex (LPHC) fractures have the highest mortality rate (27% mortality), partly due to an excessive blood loss and rupture of the pelvic structures from bony fragments. Analysis of multiple LPHC fractures mechanism, especially an influence of the dynamic forces direction and supported boundary conditions on LPHC destability will support the recovery and rehabilitation process of victims and is still under discussion.

The mechanism of a side-impact crash results in the energy transfer through the elements of a vehicle construction onto the passenger's side [2] as well as causes pelvic fractures and injuries within the hip joint due to pressure of the femur head on the acetabulum. According to the contemporary knowledge, an influence of impulse loads on the passenger's body during a traffic accident depends on velocity of a vehicle during the collision [3], and a placement of passengers in the vehicle [4], type of the vehicle, as well as age, height, and weight of the passengers [5,6]. Prediction of a pelvis fracture and understanding the stress distribution in LPHC is very problematic due to the above-mentioned factors, the complex anatomical structure of the pelvic ring and limitations of numerical methods [6,7].

The initial analysis of the pelvic fracture mechanism with the Finite Elements Method (FEM) indicates that anterior structures of the pelvis are the most susceptible to injury in the following order: right pubic ramus and ischium, next the left pubic ramus and ischium, fractions of left iliac fossa and acetabulum [8]. The bending stress due to out of plane loading was analyzed by Majmuder [9] and it was observed that the vehicle door reduced an effect of load transmission when compared to the pelvis model without the inclusion of a car door. Further studies indicate that tension concentrates mainly in the sacrum which acts as a type of a wedge against which the ilia are supported [10–12]. At the front of the pelvis, firmness and strength are provided by ligaments and the pubic symphysis. Due to this phenomena a level of elasticity is decreased and the pubic symphysis displays significantly lower tension in comparison to the sacrum. Numerous experimental studies have also indicated that the LPHC injuries depend on the side crashes: fracture of the superior pubic ramus from side impact, fracture occurring in the proximity in parallel to the pubic symphysis, and the fracture of the ischium [13–15]. The complexity and extent of LPHC injuries are also confirmed by the classification of these injuries developed by Tile, Young and Burgess [16]. Sacral bone fracture is relatively rare and difficult to evaluate (it requires accurate radiographs, hence it is often overlooked). In many cases, these types of fractures are accompanied by fractures of both pubic bones or the ischium, usually runs on the lower edge of the sacroiliac joint and bone debris moves inside the pelvis.

According to the contemporary knowledge of pelvic ring injuries, in the case of a side load, the damage to the pubic symphysis and injuries of the associated ligaments should be expected. The elaborated numerical models of a pelvic bone show that none of the studied examples attempted to determine the behaviour of the LPHC after the fracture occurred and remained under a load.

The aim of this study was to determine the mechanism of a multiple fracture formation in the pelvic ring with regard to the stress and strain in the ligaments and their role in stabilizing and energy transmission in the joint motion. This analysis also assumed the changes in cortical bone properties at the site of a comminuted fracture under closed fractures. Therefore, this model can potentially be useful for evaluating pelvis injuries in car crashes and assessing LPHC fractures and internal organ/tissue damages.

2   Materials and Methods

A detailed geometrical FE model of the LPHC was constructed from computer tomography (CT) images of an adult male after proper thresholding and segmentation with 3D Doctor and a shell-solid modelling approach with ANSYS Workbench 16.2. LPHC model included the following bones: hip, public, sacrum with the coccygeal, femoral and the adjacent ligaments. Bone structures were meshed with 8-node tetrahedral finite elements, because tetrahedral element were adopted instead of cubic or hexahedral to represent a smooth surface, which accurately depicts stress in the concentration areas with the use of local grid densification. Cancellous and cortical bones were considered as isotropic material and its mechanical properties shown in Table 1 were defined using the literature data and original studies by Bedzinski et al. [17,19,20].

2.1   Pelvis Geometric Model

All ligaments were modelled with tendinous elements (one-dimensional) with strength properties shown in Table 2. The following pelvic ligaments were integrated into the model: ligaments responsible for supporting the weight of the torso and upper limbs preventing herniation of the sacrum with the coccygeal bone in-between pelvic bones, thus reducing the vibrations transferred from the pelvis to the spine with the participation of the anterior sacroiliac ligament and the posterior sacroiliac. Additional ligaments responsible for: preventing backwards tilting of the sacrum's tip: sacrotuberous and sacrospinous ligaments were also studied.

Usually, the non-linear character of a model is accounted for by the application of both geometric and material non-linearity [18]. Material non-linearity could be included through the application of models which describe the stress-strain curve as a combination of linear functions. This approximates the behaviour of the material before and after exceeding the elasticity limit [21,22]. This is a traditional approach to bone modelling in which the model accounts for the presence of permanent deformations.

Table 1. Mechanical properties of skeletal structures.

Ligament name	Young's modulus [MPa]	Poisson's ratio
Sacrum - cortical bone	18000	0.3
Sacrum - trabecular bone	132	0.3
Pelvis - cortical bone	17000	0.3
Pelvis - trabecular bone	140	0.3
Femur - cortical bone	17000	0.3
Femur - trabecular bone	140	0.3
Cartilage sacrum-pelvis	54	0.2
Symphysis cubica	54	0.2

Table 2. Parameters for ligaments of the lumbo-pelvic-hip complex model.

Ligament name	Young's modulus MPa	Cross area mm^2	Thickness mm	Stiffness N/mm
Lig. Sacroiliaca anteriora s. ventralia	355	1.58	0.16	21000
Lig. Sacroiliaca posteriori s. dorsalna	355	1.72	0.19	18900
Lig. Sacrotuberale	355	1.28	0.21	12600
Lig. Sacrospinale	355	6.72	0.63	22500

The following assumptions describing a mutual relation between individual elements of the model were set as:

- contact between the femoral bone and the pelvis - the acetabulum was supported on the circumference of the semilunar surface,
- contact between the sacrum and the pelvis - a cartilage was modelled in the shape of the sacrum-pelvic surface (auricular surface),
- a cartilage with unique properties was applied in the pubic symphysis,
- the contact conditions between surfaces of the adjoining elements were assumed to be bounded which meant that the surfaces could not lose contact with each other or move.

The above settings describe a model in which loads are transferred not only through skeletal elements, i.e. pelvis, sacrum, or femoral bone, but also where cartilage, pubic symphysis, and the main ligaments have a significant impact on the distribution of stress.

2.2 Boundary Conditions

The impact was applied in the horizontal direction by hitting the greater trochanter of the femoral bone [23] and in the vertical direction on the anterior superior iliac spine [24] attached to several element surfaces including the lowest node of the model.

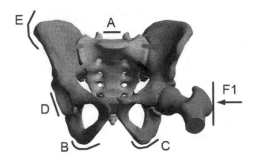

Fig. 1. Applied load (F1) and boundary conditions of the developed lumbo-pelvic-hip complex model, reflecting torso (A), seat (B, C), bucket seat (D), seat belts (E)

In order to establish the mechanism of the pelvic ring skeletal-joint system destruction under the impact loads, an influence of varying car parts on a passenger's body was determined in five cases, where the areas for setting a boundary condition were selected.

- Case I - stabilization was applied where the spine is attached which resulted in stiffening of the torso (Fig. 1A) and a consequent lack of mobility both in the thorax and the sacrum. The second support is transferred to the right ischiadic tuber which is on the opposite side of the load (Fig. 1B). This case describes a situation in which, during the side collision a person notices the approaching crash and tilts in the opposite direction.
- Case II - relates to a situation in which the upper support comes from a stiff torso (Fig. 1A), whereas the lower support takes place in the pelvic plate on the level of the femoral joint on the opposite side of the load source (Fig. 1D). Such a support can occur with the use of bucket seats or when the anterior pelvis tilt occurs.
- Case III - the support is provided to the right ischiadic tuber by the seat (Fig. 1B) and to the anterior superior iliac spine by the seat belt (Fig. 1E).
- Case IV - relates to a situation in which the lower support comes from the seat (Fig. 1C) and is provided by the muscles and the left ischiadic tuber, whereas the upper support is provided by seat belts to the right anterior superior iliac spine (Fig. 1E).
- Case V - the support is provided by the right anterior superior iliac spine (opposition of the load direction) which reflects the seat belt (Fig. 1D) and the right femoral bone (Fig. 1E).

3   Results

Several simulations with the lumbo-pelvic-hip complex (LPHC) model were performed to determine the mechanism of pelvic injury and to predict its formations. The elaborated model and cases I–V described in the material and method section were used to analyze the impactor velocity, and an impactor with or without padding on the biomechanical responses and pelvis injuries.

3.1 Force Magnitude

A magnitude of the contact force between the body and the vehicle parts during a side collision were determined by the driver's door in elaborated model. A fracture threshold was established by applying lateral loads from 0 to 10 kN to the greater trochanter of the femoral bone in each of five cases of boundary conditions. The limit of the force causing the fracture was determined based on the maximum stress and was equal to 160 MPa. Using this method the experiment established the order of damage occurrence in skeletal structures and also determined the force which causes the bone fracture (Fig. 2). The highest stress values were recorded in cases I and II which are related to situations where the sacrum base is fixed. In such cases, the stress is concentrated around the sacral promontory causing thereby stress of values reaching 160 MPa under the loads of 3.5 kN for I case, and 6.5 kN for case II. Cases III and IV assume that torso and therefore the base of the sacral bone are mobile and can move freely. Consequently, stress is concentrated in the pelvic plates. For case III, the maximum stress values occur on the right pelvic bone in the inferior pubic ramus, whereas for case IV in the left pelvic bone and in the inferior pubic ramus. Exceeding 160 MPa occurs with the load force under 8 kN in both cases.

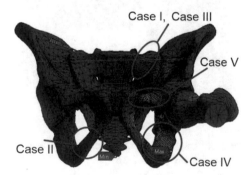

Fig. 2. The von Mises stress distribution for individual cases I–V varying by boundary conditions

3.2 Damage Proccess

Medical literature presents only a few attempts to model the pelvis after a loss of continuity in the ring structure. An example of such an analysis could be the work of Varga et al. [25]. The authors studied the stability of the pelvis with discontinued pubic symphysis with an emphasis on an importance of sacrospinous and sacrotuberous ligaments. For the purpose of rigidity changes analysis in the pelvic ring after fracture of a certain bone fragment, the researchers prepared 3 cases of the most common fractures: fracture of the sacrum side (Fig. 3A), fracture of the superior pubic ramus (Fig. 3B), and fracture of the ischium (Fig. 3C). Additionally, it was assumed that the fracture occurs within a layer of the cortical bone. This assumption was applied since the analysis of the parameters of the cortical and the cancellous bone revealed that stress concentrates on the surface (Fig. 3D, E). Changes in the bone properties were introduced while maintaining of the model continuity. It was assumed that the dislocation does not occur only in the case of a fracture at the surface of the bone because the stress values in the cancellous bone are negligible and this structure is not subjected to damage. The fragments of the model shown in Fig. 3D, E are assigned to the cancellous bone properties which allowed the cross-sectional structure of the bone to remain unaltered. Such an approach allows observation of changes in the system's rigidity without changing the bone parameters during simulation. It also eliminates a need to conduct a mathematically complex process of the bone fracture analysis. Since a fracture without dislocation occurs where the bone continuity is not interrupted in the majority of real-world cases, this approach is assumed to be accurate (Table 3).

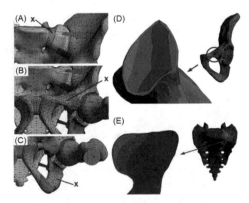

Fig. 3. The view of the bone fragment with modified properties resembling characteristics of a broken bone: model for cases I and II (A), model for cases IV and V (B), model for case III (C) and stress concentration in the cortical bone: pubic bone (D), lateral mass of the sacrum (E)

3.3   Clinical Case

This article presents a method for evaluating the effect of damaging loads impacting skeletal structures. Preliminary validation of the LPHC model was carried out using the literature data [8,9,12]. The results were obtained in accordance with the above analysis and the classification of pelvic injuries developed by Tile, Young and Burgess (Fig. 2), which confirms the correctness of the elaborated model. Furthermore, this method can be useful in the assessment of the direct injury caused to patients who suffered multiple fractures of the pelvic ring. Clinical case: a 25-year-old patient injured in a road accident (side collision), diagnosed with the fractures of the stem of the left pubic bone, the left

Table 3. Maximum von Mises stress values [MPa] in the pelvic soft tissue under the impact load of 10 kN

Case type	Breaking	Pubic symphysis	Right sacroiliac	Left sacroiliac	Acetabular fossa
I	Before	74	88	138	119
	After	120	81	70	123
II	Before	43	9	74	124
	After	42	9	65	128
III	Before	76	86	56	123
	After	74	76	61	127
IV	Before	25	34	81	103
	After	26	34	71	116
V	Before	55	54	37	126
	After	54	50	33	130

superior pubic ramus, the left ischium ramus, and the left lateral mass of the sacrum (Fig. 4).

It needs to be assumed, therefore, that in the case where a vehicle is moves in the normal direction to the driver's door at the moment of collision (with obstacle) that the fictitious force impacting the body causes it to transfer its weight onto the ischium on the side of the force. In such a situation, the stabilization of the pelvis resembles case IV and, therefore, a fracture of the ischium occurs. A Significant displacement caused by door deformation resulted in the body weight being transferred onto the right ischium thus changing the boundary conditions as in case V. As a result the fracture of the superior pubic ramus occurs. Airbags caused the immobilised torso of the driver to change the boundary conditions to case II. This resulted in stress concentration in the lateral mass of the sacrum.

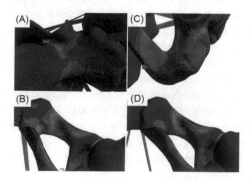

Fig. 4. Stress concentration in the lumbo-pelvic-hip complex on the load side: case I and II (A), case III (B), case IV (C), case V (D)

None of the previously presented boundary conditions clearly define the mechanism of such an injury. Therefore, it should be assumed that these injuries occurred as a result of body position changes and, subsequently, the changes in boundary conditions during the collision (Fig. 5).

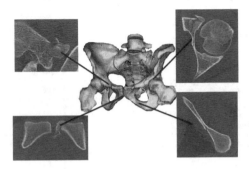

Fig. 5. The clinical case of a patient with extensive fractures within the pelvic ring on the basis of CT image

4   Discussion

The objective of this work was to analyze the mechanism of multiple pelvis fractures in side-impact car crashes. The elaborated numerical model of the lumbo-pelvic-hip complex includes the most important details responsible for its rigidity and strength including a pelvic, hip, and sacral bones as well as soft tissues such as ligaments and cartilages. This analysis assumes that bone fractures without dislocation cause a change in rigidity of the whole skeletal system. Each of five sets of boundary conditions was used to conduct the numerical experiment where a horizontal load range 0–10 kN was applied. Using this method, the experiment established the order of damage occurrence in skeletal structures and also determined the force which causes bone fracture. The obtained results correspond to the previous research and the linear correlation between static loads and von Misses stress elaborated by El-Asfoury et al. [8, 12] established the value of force which causes pelvis destruction during the side collision to be 8.6 kN. Majumder et al. [9] indicated that the maximum von Mises stress for the individual areas, i.e. the pubic symphysis, exceeds the compressive strength (200 MPa) under the load of 5 kN. The analysis provided in this paper contributes to determine an influence of boundary conditions on pelvic fractures. The above stress distribution analysis shows that the points of maximum stress significantly depending on the boundary method of the model. The performed analysis also established which fracture occurs first.

The issue of modelling the skeletal system when subjected to loads causing multi-point comminuted fractures of the pelvic ring was limited to a linear issue. It was possible because the material is subjected to pressure (load) to the point when structural deformations appear. The presented methodology involves alteration of the system rigidity resulting from a change in mechanical properties of the selected fragments of skeletal structures after exceeding this threshold. Bone fragments, where properties were altered, were selected on the basis of stress concentrations analysis which suggests the occurrence of permanent deformations, cracks, and fractions. The presented analysis assumes that the occurring fractures are of a closed type without a loss in material continuity. Fracture occurrence in the cortical bone was modelled by changing the material's properties.

These models can be successfully applied in cases where the analysis is concerned with the system behaviour from the beginning of loading until the elasticity threshold is exceeded. Exceeding the elasticity threshold of the bone should be understood as the equivalent of permanent damage and, therefore, a change in mechanical properties of the damaged area. In the case of multiple fracture, a plastic or viscoelastic material model is not sufficient due to a lack of ability to include changes in the properties of damaged tissues. Moreover, such models do not include the impact of these damages on the changes in stress distribution in other bone structures and tissues.

A similar approach to pelvis modelling was not found in literature. In previous studies researchers concentrated mainly on the stress of certain structures and on indicating the concentration points of the highest stress as well as the points where the assumed deformation threshold was exceeded.

Due to the fact that multi-point and comminuted fractures are relatively common in clinical practice, approaches proposed by previous studies [26–29] appear to be insufficient for clear description of the mechanism of multi-point injuries. The authors chose a clinical case with multi-point and comminuted fractions of the pelvic ring resulting from a road accident as an examplary calculation. The patient described in this work suffered the injuries during a side collision of the vehicle, and the fractions occurred both in the pelvic plate as well as in the sacrum. The approach proposed in this work allowed the model to recreate the order of fraction occurrence. Good agreements were found between the simulation results of the model and the experimental data in the contact force, sternum displacement, and force-displacement response. These data suggest that this 3D FE model is effective and suitable for assessing pelvis biomechanical responses upon the impact loading. Therefore, it was assumed that this fractured structure has properties similar to a cancellous bone.

5 Conclusions

The study of tissue structure destruction allows the development of an effective method to determine the mechanism and the occurred fractures. This analysis demonstrates a significant correlation between the position of the body in the car and boundary conditions. These conditions therefore have an impact on the distribution of stress and the points at which it is concentrated.

148   T. Klekiel et al.

Acknowledgement. The research was done within the project no. DOBR-BIO4/022/13149/2013 'Improving the Safety and Protection of Soldiers on Missions Through Research and Development in Military Medical and Technical Areas' supported and co-financed by NCR&D, Poland.

References

1. Lopez-Valdes, F.J., Lau, S.H., Riley, P.O., Lessley, D.J., Arbogast, K.B., Seacrist, T., Balasubramanian, S., Maltese, M
2. Kent, R.: The six degrees of freedom motion of the human head, spine, and pelvis in a frontal impact. Traffic Inj. Prev. **15**(3), 294–301 (2014)
3. Klekiel, T.: Biomechanical analysis of lower limb of soldiers in vehicle under high dynamic load from blast event. Ser. Biomech. **29**(2–3), 14–30 (2015)
4. Klekiel, T., Bedzinski, R.: Finite element analysis of large deformation of articular cartilage in upper ankle joint of occupant in military vehicles during explosion. Arch. Metall. Mater. **60**(3), 2115–2121 (2015)
5. Beason, D.P., Dakin, G.J., Lopez, R.R., Alonson, J.E., Bandak, F.A., Eberhardt, A.W.: Bone mineral density correlates with fracture load in experimental side impacts of the pelvis. J. Biomech. **36**, 219–227 (2003)
6. Rowe, A.S.: Pelvic ring fractures: implications of vehicle design, crash type, and occupant characteristic. Surgery **136**(4), 842–847 (2004)
7. Etheridge, B.S., Beason, D.P., Lopez, R.R., Alonso, J.E., McGwin, G., Eberhardt, A.W.: Effects of trochanteric soft tissues and bone density on fracture of the female pelvis in experimental side impacts. Ann. Biomed. Eng. **33**(2), 248–254 (2005)
8. Dawson, J.M., Khmelniker, B.V., McAndrew, M.P.: Analysis of the structural behavior of the pelvis during lateral impact using the finite element method. Accid. Anal. Prev. **31**, 109–119 (1999)
9. Majumder, S., Roychowdhury, A., Pal, S.: A finite element study on the behavior of human pelvis under impact through car door. In: Paper Presented at: 1st International Conference on ESAR, Hannover, German (2004)
10. Li, Z., Kim, J., Davidson, J.S., Etheridge, B.S., Alonso, J.E., Eberhardt, A.W.: Biomechanical response of the pubic symphysis in lateral pelvic impacts: a finite element study. J. Biomech. **40**, 2758–2766 (2007)
11. Majumder, S., Roychoowdhury, A., Pal, S.: Dynamic response of the pelvis under side impact load - a three-dimensional finite element approach. Int. J. Crashworthiness **9**(1), 89–103 (2004b)
12. El-Asfoury, M.S.: Static and dynamic three-dimensional finite element analysis of pelvic bone. Int. J. Math. Phys. Eng. Sci. **3**(1), 36–41 (2009)
13. Plummer, J.W., Eberhardt, A.W., Alonso, J.E., Mann, K.A.: Parametric tests of the human pelvis: the influence of load rate and boundary condition on peak stress location during simulated side impact. Adv. Bioeng. **39**, 165–166 (1998)
14. Sarlak, A.Y.: An unusual type of lateral compression injury of the pelvis tilt fracture with anterior displacement. Injury **40**, 1036–1039 (2008)
15. Yoganandan, N., Pintar, F.A., Stemper, B.D., Gennarelli, T.A., Weigelt, J.A.: Biomechanics of side impact: injury criteria, aging occupants, and airbag technology. J. Biomech. **40**(2), 227–243 (2007)
16. Rommens, P.M., Hofmann, A.: Comprehensive classification of fragility fractures of the pelvic ring: recommendations for surgical treatment. Injury **44**, 1733–1744 (2013)

17. Bedzinski, R., Nikodem, A.M., Ścigała, K., Dragan, S.: Mechanical and structural anisotropy of human cancellous femur bone. J. Vibroeng. **11**(3), 571–576 (2009)
18. Shim, V., Bohme, J., Vaitl, P., Klima, S., Josten, C., Anderson, I.: Finite element analysis of acetabular fractures - development and validation with a synthetic pelvis. J. Biomech. **43**, 1635–1639 (2010)
19. Bedzinski, R., Wysocki, M., Kobus, K., Szotek, S., Kobielarz, M., Kuropka, P.: Biomechanical effect of rapid mucoperiosteal palatal tissue expansion with the use of osmotic expanders. J. Biomech. **44**(7), 1313–1320 (2011)
20. Bedzinski, R.: Selected problem in application of experimental and numerical methods in the biomedical engineering. In: Paper presented at: 27th Danubia-Adria Symposium on Advances in Experimental Mechanics, Wrocław, Poland (2010)
21. Keyak, J.H., Rossi, S.A., Jones, K.A., Skinner, H.B.: Prediction of femoral fracture load using automated finite element modeling. J. Biomech. **31**, 125–133 (1998)
22. Keyak, J.H.: Improved prediction of proximal femoral fracture load using nonlinear finite element models. Med. Eng. Phys. **23**, 165–173 (2001)
23. Bessho, M., Ohnishi, I., Matsumoto, T., Ohashi, S., Matsuyama, J., Tobita, K., Kaneko, M., Nakamura, K.: Prediction of proximal femur strength using a CT-based nonlinear finite element method: differences in predicted fracture load and site with changing load and boundary conditions. Bone **45**, 226–231 (2009)
24. Bekker, A., Kok, S., Cloete, T.J., Nurick, G.N.: Introducing objective power law rate dependence into a viscoelastic material model of bovine cortical bone. Int. J. Imp. Eng. **66**, 28–36 (2014)
25. Varga, E., Balázs, E.: Severe pelvic bleeding: the role of primary internal fixation. Eur. J. Trauma Emerg. Surg. **36**(2), 107–116 (2010)
26. Golman, A.J., Danelson, K.A., Miller, L.E., Stitzel, J.D.: Injury prediction in a side impact crash using human body model simulation. Accid. Anal. Prev. **64**, 1–8 (2014)
27. Ma, Z., Lan, F., Chen, J., Liu, W.: Finite element study of human pelvis model in side impact for Chinese adult occupants. Traffic Inj. Prev. **16**(4), 409–17 (2015)
28. Schiff, M.A.: Risk factors for pelvic fractures in lateral impact motor vehicle crashes. Accid. Anal. Prev. **40**, 387–391 (2008)
29. Viano, D.C., Lau, I.V., Asbury, C., King, A.I., Begeman, P.: Biomechanics of the human chest, abdomen and pelvis in lateral impact. Accid. Anal. Prev. **21**, 553–574 (1989)

Analysis of the Lower Limb Model Response Under Impact Load

Tomasz Klekiel[1](✉), Grzegorz Sławiński[2], and Romuald Będziński[1]

[1] Biomedical Engineering Division, University of Zielona Gora,
Licealna 9 Street, 65-547 Zielona Gora, Poland
t.klekiel@ibem.uz.zgora.pl
[2] Faculty of Mechanical Engineering, Department of Mechanics and Applied
Computer Science, Military University of Technology,
Gen. Witolda Urbanowicza 2 Street, 00-908 Warsaw, Poland

Abstract. The paper presents a problem of soldiers' lower limbs safety in military vehicles during high impact loads derived from explosion of Improvised Explosion Devices (IED) charges. The numerical studies concerned the function of combat boots as an element of the soldier's equipment. The model of a lower limb with cooperation with a sole was prepared as a multibody dynamic system. For this model the governing equations have been prepared and solved numerically with the Runge-Kutta method. The results were obtained for the load acting on the sole as the velocity generated proportionally in relation to the mass of an IED charge. The changes of material property for the sole were analysed to select the best parameters for protection of both a foot and a whole leg. The results show the conditions for increasing the safety of a passenger's feet formulated as damping properties for soles and were compared with the similar data from an experimental study.

Keywords: Foot model · Spring · Damping · Mass · Impact load
Tissue properties

1 Introduction

In all military conflicts, a landmine technology is improved,however a group of explosive charges presenting a special role is IED which disables and destroys vehicles, injuring or killing the occupants at the same time [2,6,20]. Among all casualties, about 85% had lower limb injuries. The foot is the first part of the body having contact with the blast propagating from the floor [8]. The average velocity and acceleration of the floorplate may exceed $12 \, \text{m/s}$ with acceleration of even 100 g [21]. The dynamic response of body tissues generated from short-duration axial loads transmitted with a high amplitude leads to the damages in lower legs approximately in 10 ms [12]. The axial force generated in the Tibia is very large in comparison to other elements of the body during explosion under a vehicle [22].

© Springer Nature Switzerland AG 2019
K. Arkusz et al. (Eds.): BIOMECHANICS 2018, AISC 831, pp. 150–162, 2019.
https://doi.org/10.1007/978-3-319-97286-2_14

The lower limb is the primary impact point of an occupant. The combat boot may decrease the peak tibia axial force up to 50%, as well as increase the time-to-peak [1, 7]. The dynamic models are composed of the mass, stiffness and damping components and they create the serial, parallel and combination of both structures. These systems are developed to study the immunity of a human body or its parts during impact collisions with objects such as walls, floors or other objects in the motion. As a result, the force impulse generates a significant tissue deformation that can lead to damage [17].

For these systems, the individual masses and characteristics of stiffness and damping correspond to the physical and mechanical properties of the soft tissues, such as ligament or tendon but also the bones and cartilages. The Spring-Damping-Mass Systems (SDMS) can be represented as tissues in a variety of configurations, including one- or multi-layer structures (Fig. 1.).

The governing equations describe an energy function in the model and include information about tissue characteristics and about the environment or objects which interacted with the tissues such as shoes, clothing items, helmets, glove elements, etc. There are known many methods used to evaluate the structure stiffness, however all of them are directly applicable to the assessment of the limb stiffness. Butler [3] studied stiffness methods used to approximate axial rigidity of the lower limb in a one-dimensional system with the mass and stiffness. In general, there are some approaches in solving the problem in which a simple proportional relation between the force and deformation was assumed [13]. The other method developed by Cavagna [4] assumes that stiffness is expressed as a sinusoidal wave motion by the mass of the system multiplied by the second power of velocity. Other relations are defined by the vibration frequency.

There are many modified versions of models with the structure shown in Fig. 1. The version of the model proposed by Yue and Mester was used to study the body response to vibration acting on the ground where the effects of the body part vibrations and movable tissues have an influence on reduction the impact forces between the body rigid segments [23].

The human body model developed by Ly et al. [11] was a new version of the model proposed by Nigg and Liu [16]. In this model, the strength between a foot and a sole was defined by the non-linear stiffness functions. The authors suggested that the characteristic of the sole material does not significantly reduce the peak load. On the other hand, they suggested a good correlation to decrease the peak loading by mechanical properties of the ground. According to the calculations presented in this paper, on influence of the sole material properties on reducing the force generated in the lower limb is studied.

The aim of the investigation was focused on an influence of the combat boot sole characteristic on the tissues stress in the foot. The forces generated in the Tibia bone were compared with the similar results from other investigations. For the high impact loading generated from IED accidents, the boot sole can reduce a significant part of the energy transmitted to the lower limb. The material properties of the sole for a good protection of lower limb were proposed as a result of the investigations presented in this paper. The explosion involves large deformations with velocities unusual for typical mechanical systems. The energy is transmitted through mechanical elements in the direction of wave propagation. If the wave forehead encounters an obstacle the pressure can generate the acceleration even above 100 g. The internal energy cumulated in the material is very high and needs to dissipated as to protect the tissue against damage. A similar problem occurs in the case of stability traction in vehicles [10]. The energy dissipation can be realised by vibrating movement of the obstacle, however the main role of protection is realized by the boots. In this paper, the reaction of a muscle in the lower limb as the vibration object was investigated.

2   Materials and Methods

A mathematical model of the lower limb with a boot sole is based on the movement equations describing the relation between the kinematic parameters of the system. The Lagrange Eq. (1) is the most popular form of the energy equation describing the system of the moving bodies:

$$\frac{d}{dt}\left(\frac{\partial V}{\partial \dot{x}_k}\right) + \frac{\partial P}{\partial x_k} + \frac{\partial D}{\partial x_k} = F(t) \tag{1}$$

Based on Eq. (1), the set of equations defined the SDMS presented in Fig. 1 and was calculated with assumption that the energy occurred in the kinetic, potential and dissipation form Eq. (2) described the kinetic energy V which has the following form:

$$\frac{d}{dt}\left(\frac{\partial V}{\partial \dot{x}_k}\right) = \sum_{i=1}^{n} m_i \sum_{j=1}^{i} \dot{x}_k \tag{2}$$

The general potential energy P (3) is different and depends on the stiffness coefficient which can depend or not depend on deformation:

$$\frac{\partial P}{\partial x_k} = \frac{\partial}{\partial x_k}\left(\frac{1}{2}\sum_{i=1}^{n} k_i x_i^2\right) = \frac{1}{2}x_k\left(\frac{\partial}{\partial x_k}\left[k_k(x_k)\right] + 2k_k(x_k)\right) \tag{3}$$

The dissipation energy D is described by relation (4). Usually, the damping coefficient has a constant value, but it can also depend on deformation and/or velocity of deformation. For viscoelastic materials the damping coefficient usually depends on velocity of deformation.

$$\frac{\partial D}{\partial \dot{x}_k} = \frac{\partial}{\partial \dot{x}_k}\left(\frac{1}{2}\sum_{i=1}^{n} c_i \dot{x}_i^2\right) = \frac{1}{2}\dot{x}_k\left(\frac{\partial}{\partial \dot{x}_k}\left[c_k(\dot{x}_k)\right] + 2c_k(\dot{x}_k)\right) \tag{4}$$

The derived Eqs. (2–4) can be used in the SDMS with any number of variables. In this article, the presented equations were used to analyse the two- and three-body systems. The first model concerns the behaviour of the shoe sole-limb under impact of the axial load acting one the sole perpendicular to the ground.

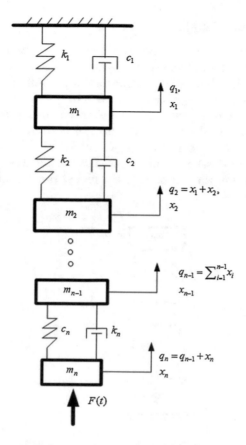

Fig. 1. General model of the Spring-Damping-Mass System (SDMS)

As a result, the model includes the conditions for effective energy dissipation by the shoe soles under high loading. The second model involves vibrations of the soft tissue mass. The set of equations was solved with the Runge-Kutta method implemented in the script prepared in the SCILAB software. The calculation was realised for a general force F_i defined as an impulse of the force generated from the explosion of an IED charge under the vehicle added to the model as F_i. For the next segments the force value equalled zero.

3   Results

The values of the individual parameters of the SDMS are an important factor for finding a solution. Usually, the parameters are taken by the numerical analysis, estimation or most often by experiments. In this paper, the parameters were selected based on Nikooyan work [17].

3.1   The Sole-Limb Model

The model of the shoe sole is composed of two masses connected by one spring and one damping element. Mass m_1 is an equivalent of the shoe weight. The factors k_1 and c_1 describe the material properties of the sole stiffness and damping, respectively. The mass m_2 represents the mass of the lower leg. The stiffness k_2 and damping c_2 parameters are equivalent to mechanical properties of the lower limb. The model has two degrees of freedom and both masses are moving in one vertical direction. The system has two general variables as the general coordinates correspond to masses m_1 and m_2, respectively. These general coordinates describe the position of these masses in the global static coordinate system.

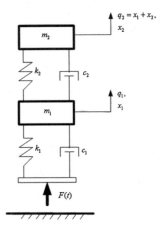

Fig. 2. The model of the set sole-limb

The force impulse was formulated with the maximum value of 12 kN and the pulse duration equalled 0.2 s. This waveform is a simplified form of the waveforms recorded on the floor during explosion under the vehicle. A reference to the anthropometric values of the body weight was used to simpler determination of the masses. The values of the stiffness and damping coefficients can be empirically determined by measuring the acceleration in the impact test and then identifying the coefficients by the equation according to these data. The stiffness of some bodies can be determined based on mechanical parameters, such as the Young's modulus or transverse stiffness expressed by the Kirchhoff modulus in reference to the mean cross-sectional area. Assuming that the contact surface of the body is known, the stiffness coefficient is used to define how the body is deformed by the force. A material can be characterized by a variety of the stiffness depending on a deformation rate. This dependence needs to taken into account by using any mathematical formula. For example, a polynomial equation may be used to determine this relation.

The stiffness coefficients can also be determined by the numerical techniques such as a finite element method. The relationship between deformation and force

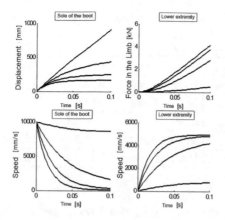

Fig. 3. The results of movement for force impulse equal to 12 kN and different values of sole damping coefficient 0.5, 10, 20, 30

is definable when the load conditions of a given element are known. This method can be applied to relatively simple systems with non-complicated shapes and structures due to computational complexity. Determination of nonlinear characteristics may result from adoption of a nonlinear types of the material model such as viscoelastic. The stiffness and damping functions can also take into account the effects of the total compression or loss of continuity. In the literature, several cases of empirical and numerical research describe that a dangerous load rate acting on the leg is in the range from 7 m/s to 10 m/s. For this rate of the floor deformation, the load force generated on the leg is up to 10–14 kN [9,15,18,19]. The loading of the leg is generated as a result of the floor deformation under the explosion and blast loading [15]. In practice, the force generated in the leg is measured by the force sensors placed in the dummy leg [19] or acceleration sensors placed on the leg. For designation of the loading, a change of speed during motion can be used to determine the bone deformation. Assuming that the relation between strain and stress is known, a value of the force generated in the limb can be calculated. The results of the sole-limb model presented in Fig. 2 suggested that the limb deformation was adopted on the basis of the assumed stiffness coefficient k_2 including the limb as a whole. The force generated during the load is determined based on a k_2 coefficient. For the established relation between the impact force rate and value of the force generated in the Tibia, a velocity of the floor was selected where the value of initial velocity of 10 m/s in time t = 0 s was assumed. Figure 3 shows the results of the experiment for different values of the sole damping coefficient. As a result, for the damping sole factor 30 Ns/mm2 the speed of the leg is minimised and equals 650 mm/s. The rate of the sole deformation suggested (Fig. 3 left-down) high convergence with the minimal amplitude. The results were obtained for the model parameters shown in Table 1. The analysis for the maximal value of the shoe damping coefficient caused that the actual deformation velocity of the sole was reduced from 5 m/s

Table 1. The parameters for the sole-limb model.

Sign	Name	Value	Unit
m_1	Mass of sole	0.5	kg
m_2	Mass of lower limb	10	kg
k_1	Sole stiffness	15	N/mm
k_2	Limb stiffness	0.86	N/mm
c_1	Sole damping	0.5–30.5	Ns/mm
c_2	Limb damping	0.94	Ns/mm

to about 0.5 m/s. At the same time, the force generated in the Tibia decreased almost eight times, from 4 kN to 0.5 kN.

3.2   The Sole-Limb Model with Vibration of the Soft Tissue

Figure 4 shows a model of the sole-limb system enriched with the mass m_3, which represents the mass of a soft tissue, mainly the muscles which move relative to the limb [24]. The problem of the muscle mass movement and other tissues during gait and running regards the generation of the maximal forces inside the limb [17]. The Fig. 4 shows the sole-limb model which includes the vibrations of the muscle mass. The sole velocity at time t = 0 s, as an initial condition, was assumed as 10 m/s in the vertical up direction. Sensitivity of the sole stiffness for the system stiffness was studied. In Fig. 5, the results are presented in the form of 6 graphs. The top graphs represent a change of displacements for a sole,

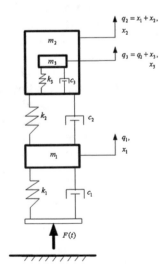

Fig. 4. The model of the sole-limb set with muscle mass vibration

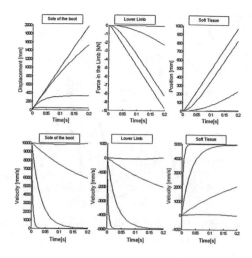

Fig. 5. The results of movement for the sole-limb model for different values of a sole damping coefficient

limb and a soft tissue. Bottom graphs show the velocity changes. The system was tested for the velocity variations, seeking the situation in which velocity will be convergent to the minimum in the shortest time possible. The optimal value of the damping factor was sought as a result of the subsequent experiments. The optimal criterion was defined based on the relations between the value of damping coefficient and the limb speed value for which the existing forces will not exceed the permissible values. The little stiffness of the shoe sole resulted in a significant reduction of the forces occurring in the limb. The numerical experiment presented in this paper demonstrated that the muscle movements do not significantly influence the response of the whole system, thereby increasing or decreasing the forces generated inside the limb. Figure 5 shows that the high

Table 2. The parameters for the sole-limb model with vibration.

Sign	Name	Value	Unit
m_1	Mass of sole	0.5	kg
m_2	Mass of lower limb	10	kg
m_3	Mass of muscle	5	kg
k_1	Sole stiffness	15	N/mm
k_2	Limb stiffness	0.86	N/mm
k_3	Muscle stiffness	0.1	N/mm
c_1	Sole damping	0.15 1.5 15 150	Ns/mm
c_2	Limb damping	0.94	Ns/mm
c_3	Soft tissue damping	0	Ns/mm

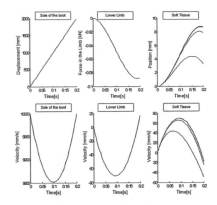

Fig. 6. The results of movement for the sole-limb model for different values of a muscle stiffness coefficient

value of the damping coefficient (damping of 150 Ns/mm2) with a relatively soft sole (stiffness coefficient equal 15 N/mm), allows damping the pulse energy leading to a situation in which, after 50 ms, the velocity of both the sole and the limb is minimal. Next, for the selected values of the damping property for the sole, the experiment focused on a relation between the acting force, defined as its stiffness, and behaviour of the whole system. Figure 6 shows the graphs of variations in deformation of the system under the influence of the initial velocity of 10 m/s. The changes of the stiffness values were assumed as 2, 20 and 200 N*s/mm. These values were selected numerically to model three main muscle states: without acting, acting normally and highly strong activated. The analysis of the effect of the muscle rigid change was performed by the stiffness parameters k_3 imitating the situations of tight and loose muscles suggesting that when the muscle is loose, its stiffness is small and muscle mass can move freely along the direction of the movement with quite high amplitude. However, when the

Table 3. The parameters for the sole-limb model with a vibration effect.

Sign	Name	Value	Unit
m_1	Mass of sole	0.5	kg
m_2	Mass of lower limb	10	kg
m_3	Mass of muscle	5	kg
k_1	Sole stiffness	15	N/mm
k_2	Limb stiffness	0.86	N/mm
k_3	Muscle stiffness	2 20 200	N/mm
c_1	Sole damping	0.15	Ns/mm
c_2	Limb damping	0.94	Ns/mm
c_3	Soft tissue damping	0.1	Ns/mm

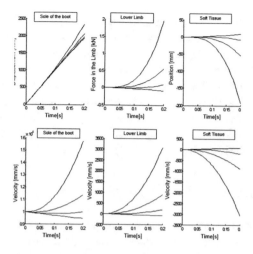

Fig. 7. The results of movement for the sole-limb model for different values of a sole stiffness coefficient

muscle are acting, the movement is limited by the change in muscle stiffness. As a result, the non-free movement of a tight muscle closely depends on the tendon and tendon deformation characteristics of the muscle. The system has been tested for various muscle stiffness simulating an acting and a loose muscle. As a result, similar changes in the position of the soles and limbs were observed in Fig. 5. It suggests that the movement of the muscle of the lower leg does not affect the behaviour of the system as a whole. There was no a significant effect on the level of deformities and loads of the bone structures in the limb. The changes involved only the muscle movements. The amplitude of the movement for loose muscle was about 4 mm but for the acting was about 8 mm. Next experiment focused on relations between stiffness of the shoe sole and forces generated in

Table 4. The parameters for the sole-limb model with a vibration effect.

Sign	Name	Value	Unit
m_1	Mass of sole	0.5	kg
m_2	Mass of lower limb	10	kg
m_3	Mass of muscle	5	kg
k_1	Sole stiffness	0.15 1.5 5 15	N/mm
k_2	Limb stiffness	0.86	N/mm
k_3	Muscle stiffness	0.1	N/mm
c_1	Sole damping	0.15	Ns/mm
c_2	Limb damping	0.94	Ns/mm
c_3	Soft tissue damping	0.1	Ns/mm

the limb. Figure 7 shows the changes during loading for different values of the sole stiffness. For the harder materials, the force generated in both the sole and the limb was substantially greater. The results presented in Fig. 7 show that the low stiffness of the sole well affects the soft tissues, significantly reducing the deformations in the limb. The lower stress in the muscle contributes to alleviating the pain caused by the overload of the tissue structures with a sudden impulse of the force. Smaller sole stiffness leads a higher deceleration, occured by a quickly convergence of the velocity to the zero. A smaller deformation shows a smaller forces in the lower limb (Table 2, 3 and 4).

4   Conclusions

The effects presented in this article are very important because they concern the differences in mechanical properties of biological tissues and its influence on the response generated by the force impulse. The fact that the endurance level of the biological structures of the human movement system is quite varied leads to conclusion that the relationship between biological properties and the system explains the conditions during designing a new protection system for soldiers. Soft tissues have a natural ability to absorb energy. The collagen fibers included within the structure are straightened during stretching. As a result, the stresses are increased nonlinearly with increasing elongation and this phenomena dissipates a significant part of energy. Initially, the tension increases to a small extent. During larger deformations, when the collagen fibers are straightened, the stiffness of the tissue is also increased. The mechanical strength of the tissues depends on i.a.: topography, age, chemical and physical aspects. These all aspects play an important role and lead to the large data spans. However, as demonstrated in the numerical experiments, the level of the force from these phenomena in mus-cles is much smaller than the forces generated during an explosion. Therefore, it should be recognised that the movement of the muscle tissue mass during the impulse load is of minor importance. Based on the presented results, the optimal parameters for the boot sole were selected. The damping properties of the soft tissue does not have an important role in energy dissipation, however an the artificial barrier in the form of a sole is possible to reduce a high force impulse from the blast event.

Acknowledgments. The research was done within the project no. DOBR-BIO4/022/13149/2013 'Improving the Safety and Protection of Soldiers on Missions Through Research and Development in Military Medical and Technical Areas' supported and co-financed by NCR&D, Poland.

References

1. Barbir, A.: Validation of Lower Limb Surrogates as Injury Assessment Tools in Floor Impacts Due to Anti-vehicular Landmine Explosions. Wayne State University Biomedical Engineering, Detroit (2005)
2. Bird, R.: Protection of vehicles against landmines. J. Battlef. Technol. **4**, 14 (2001)
3. Butler, R.J., Crowell, H.P., Davis, I.M.: Lower extremity stiffness: implications for performance and injury. Clin. Biomech. **18**(6), 511–517 (2003)
4. Cavagna, G.A., Franzetti, P., Heglund, N.C., Willems, P.: The determinants of the step frequency in runni1g, trotting and hopping in man and other vertebrates. J. Physiol. **399**, 81–92 (1988)
5. Chen, W.-M., Park, J., Park, S.-B., Shim, V., Lee, T.: Role of gastrocnemius-soleus muscle in forefoot force transmission at heel rise - a 3D finite element analysis. J. Biomech. **45**, 1783–1789 (2012)
6. Harris, R.M., et al.: Lower Extremity Assessment Program (LEAP 99–2), Fort Sam Houston, Texas: U.S. Army Institute of Surgical Research, Extremity Trauma Study Branch (2000)
7. Keown, M.: Evaluation of Surrogate Legs Under Simulated AV Loads-phase IV. Biokinetics and associates Ltd., Ottawa (2006)
8. Klekiel, T.: Biomechanical analysis of lower limb of soldiers in vehicle under high dynamic load from blast event. Ser. Biomech. **29**(2–3), 14–30 (2015)
9. Klekiel, T., Bedzinski, R.: Finite element analysis of large deformation of articular cartilage in upper ankle joint of occupant in military vehicles during explosion. Arch. Metall. Mater. **60**(3), 2115–2121 (2015)
10. Kosiński, W., Oliferuk, W.: Stationary action principle for vehicle system with damping. Acta Mechanica et Automatica **6**(4), 23–26 (2012)
11. Ly, Q.H., Alaoui, A., Erlicher, S., Baly, L.: Towards a footwear design tool: influence of shoe midsole properties and ground stiffness on the impact force during running. J. Biomech. **43**(2), 310–317 (2010)
12. Mckay, B.J., Bir, C.A.: Development of a lower extremity injury criterion for military vehicle occupants involved in explosive blast events. Belgium, Brussels (2008)
13. McMahon, T.A., Cheng, G.C.: The mechanics of running: how does stiffness couple with speed? J. Biomech. **23**(Suppl. 1), 65–78 (1990)
14. Newell, N., Masouros, D.S., Ramasamy, A., Bonner, J.T., Hill, M.A., Clasper, C.J., Bull, M.A.: Use of cadavers and anthropometric test devices (ATDs) for assessing lower limb injury outcome from under-vehicle explosions. In: IRCOBI Conference (2012)
15. Newell, N., Neal, W., Pandelani, T., Reinecke, D., Proud, W.G., Masouros, S.D.: The dynamic behaviour of the floor of a surrogate vehicle under explosive blast loading. J. Mater. Sci. Res. **5**(2) (2016)
16. Nigg, B.M., Liu, W.: The effect of muscle stiffness and damping on simulated impact force peaks during running. J. Biomech. **32**(8), 849–856 (1999)
17. Nikooyan, A.A., Zadpoor, A.A.: Mass-spring-damper modelling of the human body to study running and hopping - an overview. Proc. Inst. Mech. Eng. H. **225**(12), 1121–1135 (2011)
18. Nilakantan, G., Tabiei, A.: Computational assessment of occupant injury caused by mine blasts underneath infantry vehicles. Int. J. Veh. Struct. Syst. **1**(1–3), 50–5 (2009)

19. Pandelani, T., Reinecke, J.D., Beetge, F.J.: In pursuit of vehicle landmine occupant protection: Evaluating the dynamic response characteristic of the military lower extremity leg (MiL- Lx) compared to the Hybrid III (HIII) lower leg. CSIR International Convention Centre, CSIR, Pretoria (2010)
20. Ramasamy, A., Harrisson, S.E., Clasper, J.C., Stewart, M.P.: Injuries from roadside improvised explosive devices. J. Trauma **65**, 910–914 (2008)
21. Wang, J.J., Bird, R., Swinton, B., Kristic, A.: Protection of lower limbs against floor impact in army vehicles experiencing landmine explosion. J. Battlef. Tech. **4**, 11–15 (2001). PASS 2008
22. Yoganandan, N., Pintar, F.A., Schlick, M., Humm, J.R., Voo, L., Merkle, A., Kleinberger, M.: Vertical accelerator device to apply loads simulating blast environments in the military to human surrogates. J Biomech. **48**(12), 3534–3538 (2015)
23. Yue, Z., Mester, J.: A model analysis of internal loads, energetics, and effects of wobbling mass during the whole-body vibration. J. Biomech. **35**(5), 639–647 (2002)
24. Zadpoor, A.A., Nikooyan, A.A.: Modeling muscle activity to study the effects of footwear on the impact forces and vibrations of the human body during running. J. Biomech. **43**(2), 186–193 (2010)

Numerical Analysis of the Biomechanical Factors of a Soldier Inside a Vehicle with the Pulse Load Resulting from a Side Explosion

Grzegorz Sławiński$^{(\boxtimes)}$, Piotr Malesa, and Marek Świerczewski

Military University of Technology,
Gen. Witolda Urbanowicza 2, 00-908 Warsaw, Poland
grzegorz.slawinski@wat.edu.pl

Abstract. The aim of this paper is to try to classify and assess the risk of injuries of the cervical spine during an attack on a military vehicle. In that case, the focus has been placed only on the side explosion variant, which reflects new threats observed on the basis of events from Afghanistan and not only. The risk of a threat to the life and health of the vehicle's crew increases as result of an explosion under a military vehicle. Considering that event in terms of the safety of soldiers comes down to a complex analysis of interactions between the soldier's body, seat and vehicle's structural elements. The effects of the shock wave result in interactions which cause vibrations resulting from the vibrations of the construction and the acceleration of the occupant's body.

The currently applied test conditions and criteria of the injuries of LV and LAV crew members exposed to the shock wave resulting from the explosion of the AT mine are specified in Appendix E to the NATO standard [1]. However nowadays, those requirements have been extended and included in classified documents which cannot be presented to the general public. Therefore, the assumptions resulting from the analyses of the existing cases in Afghanistan have been adopted.

This paper includes an attempt to analyse the impact of the explosion of IED on the side of a military vehicle on the risk of injuries of the cervical spine of soldiers. The analysis has been made using numerical methods in the LS-DYNA software and has been conducted considering the variable values of displacement and accelerations registered during an explosion.

Keywords: Improvised Explosive Device · Shock wave
Numerical simulations · Criterion of injuries · Cervical spine

1 Introduction

The specificity of current combat operations forces a new approach towards the threats posed by Improvised Explosive Devices (IED). According to the statistics of war victims within the last years, where the Polish armed forces were involved, the vast majority of deaths are not cause by the enemy fire but IEDs. The detonation of IEDs takes places either by the direct overturning of the wheel or their remote detonation at a

© Springer Nature Switzerland AG 2019
K. Arkusz et al. (Eds.): BIOMECHANICS 2018, AISC 831, pp. 163–176, 2019.
https://doi.org/10.1007/978-3-319-97286-2_15

Fig. 1. The KTO Rosomak vehicle after the explosion of IED [2]

small distance from a vehicle. Figure 1 presents the KTO Rosomak vehicle after the explosion of IED under the vehicle's hull.

Complex mechanisms accompanying the effects of the shock wave on soldiers inside a vehicle are the reason for injuries e.g. the injuries of lower limbs, spine and internal organs caused by vertical acceleration or head injuries caused by displacement and hitting against hard surfaces. Other dangerous injuries are the injuries of the organ of hearing or any other injuries resulting from the complexity of the effects of physical, mechanical, pressure and temperature factors as well as the consequences of those injuries.

The shock wave resulting from the detonation of an explosive under a vehicle impacts the vehicle's structure by putting it into vibrations which cause high pressure oscillations inside the vehicle. The forces generated by the vibrations of seats result in limb injuries of soldiers and spine injuries of crew members caused by hitting their heads against vehicle's sides and roof. Injuries within the head, face and neck area often lead to serious dysfunctions or even death [3].

The analysis of the statistical data of the recent military conflicts in Iraq and Afghanistan has shown that approx. 80% of all wounded soldiers suffered injuries caused by explosions and around one half of those cases resulted from the explosion of IEDs. Significant vertical loads accompanying those events, which have an impact on the soldier's body, cause the above-mentioned spine injuries [4, 5]. According to tests, fractures in the area of facet joints and bodies of the cervical spine occur as a result of the transfer of the compressive force along the bone and ligament column axis. However, tests describing injuries, the mechanisms of injuries and the kinetics of the spine during an explosion under a military vehicle are limited because the direction of force is inverted – the load is initiated from the bottom of the neck to the upper part of the spine and the head [6].

2   Statistical Analysis of the Types of Injuries of the Cervical Spine and the Head

The analysis of the data regarding military conflicts, starting from the Second World War until the last missions in Iraq and Afghanistan, has shown that the injuries of the head and neck areas constitute almost 16–30% of all injuries. The main reason for that is a more frequent use of IEDs. The situation is similar also in the case of the analysis of injuries among the soldiers of the Polish Military Contingent in Afghanistan. In that case, the main injuries of the neck area were the following: the injuries of the cervical spine structures, contusions, fractures, lividity (which indicates circulatory failure) and cervical spine pain (which may result from the injuries of the structures around that area). Contusions constituted the biggest group among all cervical spine injuries.

The comparison of that data with the available literature has shown that the percentage value of head and neck injuries in Afghanistan since 2011 is relatively similar to the literature data presented in Fig. 2 [7]. When comparing events happening on a battle-field in Afghanistan in the period between 2011 and 2014 and in 2002, it can be noticed that the percentage of injuries in the period between 2011 and 2014 is significantly higher than in 2002. It can be presumed, as the literature sources indicate, that this may be caused by a more frequent use of IEDs or modern improved body pads (Fig. 2).

The authors of the paper [8] have analysed all cases of neck injuries of British soldiers during the war in Afghanistan in the period from the beginning of January 2006 until the end of December 2010. The conducted tests have indicated that 79% of neck injuries were caused by explosions and 41% of soldiers died as a result of those injuries. Analysis showed that 85% of deaths were caused by the injuries of carotid and vertebral arteries or internal jugular veins as well as the injuries of the spinal cord.

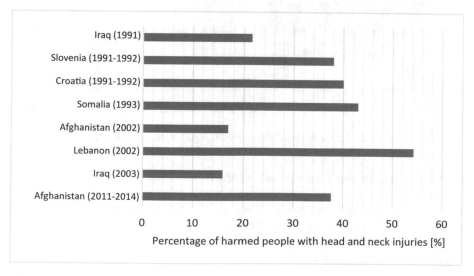

Fig. 2. The statistics of head and neck injuries on different battle-fields in given years [7]

3 Tests Regarding the Effects of the Wave of Pressure from an Explosion on a Soldier Inside a Military Vehicle

The analysis of the data from the Military Medical Institute database indicates a significant percentage of the cervical spine injuries of Polish soldiers in Afghanistan. Therefore, the aim of the tests was to try to assess the risk of injuries of the cervical spine and the head of the occupant of a military vehicle. The tests have been conducted on the basis of a dynamic numerical analysis in the LS-DYNA software, using the explicit integration method to solve issues which change fast in time.

4 Effects of the Wave of Pressure from the Detonation of an Explosive Placed on the Side of a Vehicle

The model of a vehicle in the analysed case has been loaded with an explosive, whose location is presented in Fig. 3. The centre of the explosive was located between the first and the second axis of the vehicle at the height (H) of 1.8 m from the ground, at the side distance from the transporter of 1.5 m (L).

Loading the vehicle with the wave of pressure from the detonation of the explosive with the mass of 2–40 kg has been simulated in the numerical analyses. The load has been created using the CONWEP algorithm.

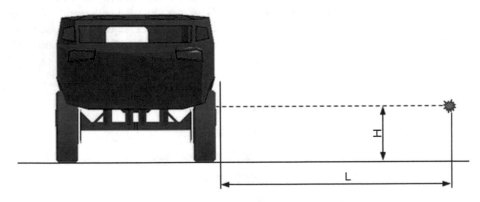

Fig. 3. The location of the explosive against the vehicle's hull

5 Structure of the Numerical Model

The EuroSID– 2re Anthropomorphic Test Device (ATD) with MIL-LX lower limbs has been used in experimental tests. That device renders it possible to measure the survivability parameters of vehicle crews during explosions and side collisions. Figure 4 presents an actual dummy.

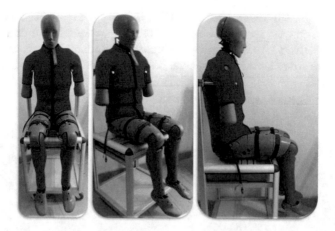

Fig. 4. The EuroSID – 2re Anthropomorphic Test Device

During the experimental test, the dummy has been placed in the tested vehicle on a seat located in its central part, on the left side, and then fastened with seat belts, which is presented in Fig. 5.

The numerical model of the EuroSID– 2re (ES-2re) dummy has been taken from the LSTC website. The ES-2re dummy is a variant of the ES-2 dummy proposed by NHTSA for the new FMVSS 214 standard. This variant has been developed to entirely delete the back plate seat interference that still may exist with the standard ES-2.

Fig. 5. The view of the EuroSID – 2re dummy inside the vehicle during field tests

The ES-2re (for Rib Extensions) is an ES-2 dummy with proposed modification to the rib unit, closing the space between the ribs and the spine. The design change is intended to effectively eliminate any grabbing of the dummy back plate into the seat structure. Figure 6 depicts the rib cages and the connection to the spine of both models.

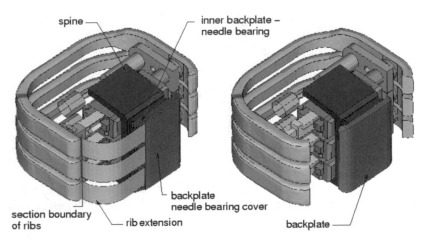

Fig. 6. The ES-2re thorax module (left), the ES-2 thorax module (right); view from the back [9]

The mass of the major parts in the model are listed in the table below (Table 1):

Table 1. The mass of the major parts of the EuroSID– 2re model

Part	Specification [kg]	Mass in kg
Head	4,0 ± 0,2	4,144
Neck	1,0 ± 0,05	1,041
Thorax	22,4 ± 1,0	21,335
Arm, left	1,3 ± 0,1	1,269
Arm, right	1,3 ± 0,1	1,38
Abdomen	5,0 ± 0,25	5,317
Pelvis	12,0 ± 0,6	11,712
Leg, left	12,7 ± 0,6	13,069
Leg, right	12,7 ± 0,6	13,069
Total	74,0 ± 1,2	72,338

The view of the numerical model of the dummy placed on the seat with visible seat belts has been presented in Fig. 7.

Fig. 7. The view of the dummy sitting on the seat with fastened seat belts

The complete numerical model of the vehicle with the dummy placed inside consists in total of 597184 finished elements. The picture below presents the view of the numerical model (Fig. 8).

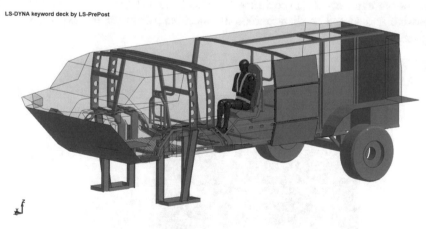

Fig. 8. The view of the numerical model of the vehicle with the visible dummy

The numerical analysis was divided into two stages. The first one lasted 100 ms and included the stabilisation of the model as well as placing the dummy on the seat by using the CONTROL_DYNAMIC_RELAXATION card, taking into consideration the effects of gravitational acceleration. During the second stage, the loading of the vehicle's body with the explosive wave was introduced. In order to simulate a TNT explosion, the Conwep algorithm was used. It is based on the empirical tests of Kingery and Bulmash, implemented in the LS-Dyna system through the Randers-Pehrson and

Bannister equation. Within that method, the value of the pressure of shock wave p striking the surface at angle θ is calculated in the following way (1):

$$p = p_i(1 + \cos \theta - 2cos^2\theta) + p_r \cdot cos^2\theta, \tag{1}$$

where:
p_i –   incident wave pressure,
p_r –   reflected wave pressure.

   The duration of the second stage of the numerical analysis was 200 ms.
   In the numerical model, 16 material models from the LS-Dyna database were used and 34 contact cards were defined. The friction coefficient of wheels and front brackets against the ground was 0.6.

6   Validation of the Numerical Model

For the purposes of the numerical model validation, the results of the field tests have been used, in which the body of the military vehicle has been loaded with a pressure wave from the detonation of 2 kg explosive. Semtex 1A was used as the explosive in that case. Its equivalent in relation to TNT is 1.25 [10]. The Conwep algorithm is based on the analysis of the propagation of pressure from a TNT explosion, therefore, the mass of the explosive assumed in the analytical calculations was 2.7 kg (TNT). The explosive was placed at characteristic distances equivalent to the experimental tests (Fig. 9).

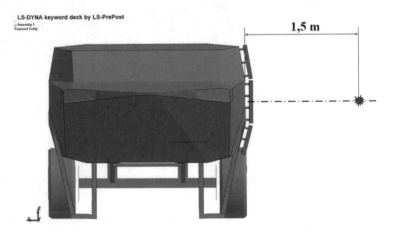

Fig. 9. The view of the vehicle with the visible location of the explosive

   The characteristics of acceleration of the dummy's head have been compared in order to validate the numerical analyses. The results from the experimental tests have been presented in Fig. 10.

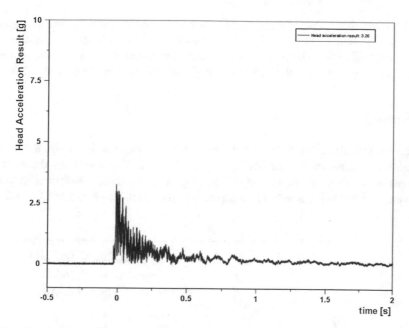

Fig. 10. The characteristic of acceleration changes of the dummy's head on the basis of the measurements from the experimental tests

Figure 11 presents the characteristic of acceleration equivalent to the above results, achieved on the basis of the experimental tests.

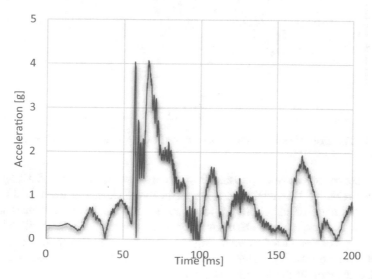

Fig. 11. The characteristic of acceleration changes of the dummy's head achieved on the basis of the numerical analyses

The maximum value of head acceleration achieved in the experimental test was 3.26 [g]. The numerical analysis gave the result of 4.07 [g], which constitutes the relative error of 25%. Taking into consideration the complexity level of the model and significant simplifications, the validation result should be considered satisfactory.

7   Results

The conducted numerical analyses have rendered it possible to learn about the impact of a side explosion on the cervical spine and the head. The results of the numerical analyses in the form of the course of bending moments, head displacement and the behaviours of the vehicle with an occupant are presented in Figs. 12, 13 and 14 and Table 2.

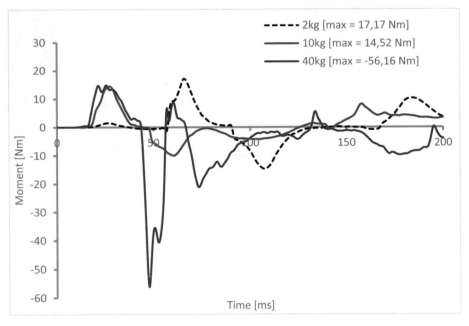

Fig. 12. The characteristic of the change of the bending moment in the dummy's neck

The effects of the wave of pressure from the explosives of 2 kg and 10 kg resulted in the occurrence of the maximum moment in the neck at the level of 14–17 Nm. Increasing the mass of the explosive to 40 kg led to significant deformations of the construction, as a result of which the dummy was hit by the vehicle's wall. The highly dynamic nature of the hit led to achieving the maximum bending moment in the neck, which was around −56 Nm. That value is close to the limit value determined in the NATO specification at the level of 57 Nm.

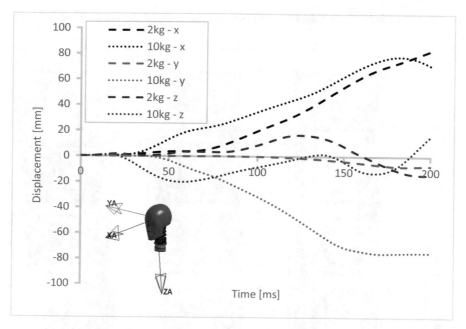

Fig. 13. The characteristic of the change of the dummy's head displacement in the case of the explosives of 2 kg and 10 kg

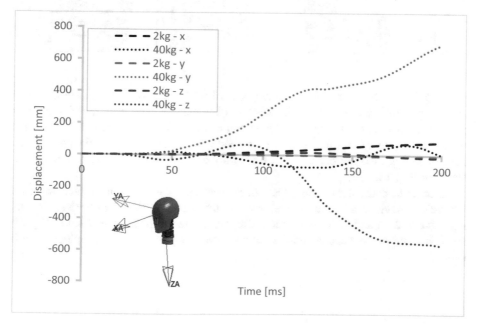

Fig. 14. The characteristic of the change of the dummy's head displacement in the case of the explosives of 2 kg and 40 kg

Table 2. Behaviour of the vehicle's construction as a result of the pressure wave from the explosion

	2kg	10kg	40kg
T=0 ms			
T=50ms			
T=100ms			
T=150ms			
T=200ms			

Also the displacement of the dummy's head in the case of the explosives of 2 kg and 10 kg were analysed. While in the case of a smaller explosive, the displacement occurs in a visible extent only in the tangential direction to the vehicle's wall (the x axis of the coordinate system in the dummy's head), in the case of an explosive with the mass of 10 kg, also the occurrence of the normal component is visible. This is presented in Fig. 13.

In order to better depict the impact of increasing the mass of the explosive on the vehicle's behaviour and thus the effects on the dummy, the characteristics of the explosive of 2 kg and 40 kg have been compared. The results are presented in Fig. 14. The behaviour of the vehicle and the dummy inside is presented in a graphical way in Table 2.

8   Summary

The essential structures of the central nervous system and the circulatory system are placed in a complex way in the neck area which, at the same time, is the most movable part of the spine. Injuries in the neck area result in serious consequences, such as permanent damage to health or death. Injured people require long convalescence and rehabilitation which involves a significant economic burden and thus completely excludes a given person from the military service.

According to the analyses of results, in the case of a soldier located in the landing area during an explosion on the side of a vehicle, the biggest displacement of the cervical spine causing sudden bending occurs in the x and y directions. The smallest displacement for loading of 2 kg TNT occurs in the z direction, where the risk of compressive spine injuries may occur. The risk of the injury of the cervical spine of the whiplash type increases in the case of the pressure impulse on the side of a vehicle. This type of injury is caused by the sudden bending of the head backwards and then its sudden bending forwards.

In order to analyse the occurrence risk of a given class of injuries, it is necessary to conduct multi-variant simulations which would include, for instance, the direction of an attack on a vehicle or a seat occupied by a soldier in that vehicle.

Acknowledgements. The research was done within project no. DOBR-BIO4/022/13149/2013 'Improving the Safety and Protection of Soldiers on Missions Through Research and Development in Military Medical and Technical Areas', supported and co-financed by NCR&D, Poland.

References

1. AEP-55, vol. 2, Edn. 1, Procedures for Evaluating the Protection Levels of Logistic and Light Armoured Vehicle Occupants for Grenade and Blast Mine Threats Level, NATO/PFP Unclassified, (2005)
2. Niezgoda, T., Slawinski, G., Gieleta, R., Swierczewski, M.: Ochrona pojazdów wojskowych przed wybuchem min i improwizowanych urzadzen wybuchowych. J. KONBiN **1**(33), 123–134 (2015)
3. Wade, A.L., Dye, J.L., Mohre, Ch.R., Galarneau, M.R.: Head, Face, and Neck Injuries During Operation Iraqi Freedom II: Results From the US Navy and Marine Corps Combat Trauma Registry. San Diego, California. Technical Report 06-01
4. Patzkowski, J.C., Blair, J.A., Schoenfeld, A.J., Lehman, R.A., Hsu, J.R.: Multiple associated injuries are common with spine fractures during war. Spine J. **12**, 791–797 (2012)
5. Mackiewicz, A., Sławiński, G., Niezgoda, T., Będziński, R.: Numerical analysis of the risk of neck injuries caused by IED explosion under the vehicle in military environments. Acta Mechanica et Automatica, vol. 10 no. 4 (2016)
6. Yoganandan, N., Stemper, B.D., Pintar, F.A., Maiman, D.J., McEntire, B.J., Chancey, V.C.: Cervical spine injury biomechanics: applications for under body blast loadings in military environments. Clin. Biomech. **28**, 602–609 (2013)
7. Rustemeyer, J., Kranz, V., Bremerich, A.: Injuries in combat from 1982–2005 with particular reference to those to the head and neck: a review. Br. J. Oral Maxillofac. Surg. **45**, 556–560 (2007)

8. Breeze, J., Midwinter, M.J., Pope, D., Porter, K., Hepper, A.E., Clasper, J.: Developmental framework to validate future designs of ballistic neck protection. Br. J. Oral Maxillofac. Surg. **51**, 47–51 (2013)
9. Schuster, P., Franz, U., Stahlschmidt, S., Pleschberger, M., Eichberger, A.: Comparison of ES-2re with ES-2 and USSID Dummy, 3. LS-DYNA Anwenderforum, Bamberg (2004)
10. Remennikov, A.: The state of the art of explosive loads characterisation. In: Lam, N., Wilson, J., Gibson, G., Anderson, S. (eds.) Australian Earthquake Engineering Conference, pp. 1–25. Australian Earthquake Engineering Society, Wollongong (2007)

Soft and Hard Tissue Biomechanics

Risk Assessment Regarding the Injuries of the Lower Limbs of the Driver of a Military Vehicle in the Case of an Explosion Under the Vehicle

Grzegorz Sławiński[✉], Marek Świerczewski, and Piotr Malesa

Military University of Technology,
Gen. Witolda Urbanowicza 2 Street, 00-908 Warsaw, Poland
grzegorz.slawinski@wat.edu.pl

Abstract. The protection of soldiers performing tactical operations in armoured vehicles plays an important role in combat operations. The need for such protection results from threats on battle-fields or during peace-keeping missions connected with the explosion of mines or Improvised Explosive Devices (IED). As a result of an explosion, the occupants of the vehicle are exposed to the effects of the shock wave as well as overloads caused by the movement of the vehicle. Simulation and experimental tests play a significant role in ensuring the safety of occupants exposed to the effects of loads caused by an explosion. It is possible to specify the conditions creating loads to the body, and also to determine interaction forces in the lower libs of the human body, using models which include the description of the explosive, parameters of the vehicle and the model of the human body.

The case connected with the behaviour of the vehicle's driver during the detonation of an explosive placed centrally under the driver's compartment has been analysed in this paper. Thanks to that, it is possible to demonstrate the influence of the seize of the explosive on the overloads of limbs. Based on the results of the calculations, it has been stated that the forces coming from the shock wave transferred to the foot of the limb are so important that they largely influence the seize of limb injuries. The reduction of the interaction force of the above-mentioned sources consists mainly in developing such a construction of the vehicle which minimises the danger of damaging the plating and renders it possible to maximally dissipate the shock wave by the vehicle's construction.

Keywords: Improvised Explosive Device · Impact load
Numerical simulations · Criterion of injuries · Lower limb

1 Introduction

The analysis of the current asymmetric military conflicts has shown that the actual threats to soldiers are largely connected with Improvised Explosive Devices (IED). After their detonation, the created shock wave causes pulse loads of the vehicle's construction. The consequences of that process may be classified in three groups: as a local effect, as a global effect (throwing the vehicle up in the air) and as a secondary

© Springer Nature Switzerland AG 2019
K. Arkusz et al. (Eds.): BIOMECHANICS 2018, AISC 831, pp. 179–193, 2019.
https://doi.org/10.1007/978-3-319-97286-2_16

effect, namely the falling down of the vehicle. The local effect occurs after the initiation of an explosion under the vehicle and the reflection of the wave from the bottom of the vehicle. As a result of that process, the floor plate bends and thus causes deformations of the side walls of the vehicle's construction. During that phase, the crew members are exposed to the strongest inertia forces. The global effect is a consequence of the reflection of the blast wave which causes the throwing of the entire vehicle up in the air and occurs after approx. 10 to 20 ms from the detonation. After reaching the maximum height determined by the mass of the vehicle and the explosive, the vehicle falls back down to the ground [1].

2   Requirements Regarding the Resistance of Military Vehicles to Explosions of Mines and IEDs

Military vehicles moving within a military conflict zone should comply with special requirements in the field of ballistic protection as well as protection against splinters, mines and IEDs. The NATO documents and the documents of institutes cooperating with NATO constitute the basis for determining requirements regarding the protection of crew members [2–4]. The protection levels of LV and LAV crew members against the shock wave from the explosion of the AT mine (Table 1) has been defined in document [2]. Levels 2–4 concern the AT land mines detonated under a wheel/track or under the centre of the vehicle. The location of the explosive under the vehicle's wheel according to the STANAG 4569 standards has been presented in Fig. 1.

Table 1. Levels of protection against the shock wave from the explosion of AT mines [2]

Level		Explosion of the AT mine	
4	4b	Explosion under the centre of the vehicle	AT 10 kg TNT
	4a	Explosion under a wheel/track	
3	3b	Explosion under the centre of the vehicle	AT 8 kg TNT
	3a	Explosion under a wheel/track	
2	2b	Explosion under the centre of the vehicle	AT 6 kg TNT
	2a	Explosion under a wheel/track	

3   Criteria Regarding the Protection of the Life and Health of Soldiers

The analysis of attacks with IEDs has shown that soldiers might suffer numerous mechanical injuries, even if the vehicle's casing is not damaged in a way which would lead to the direct penetration of the crew's compartment (with the wave of pressure or hot gases). As a result of the detonation of an explosive under the vehicle, the shock wave impacts the vehicle's structure (the bottom, sides and the roof) by putting it into vibrations which contribute to the occurrence of high pressure oscillations inside the vehicle. The forces generated by the vibrations of sets result in limb injuries of soldiers

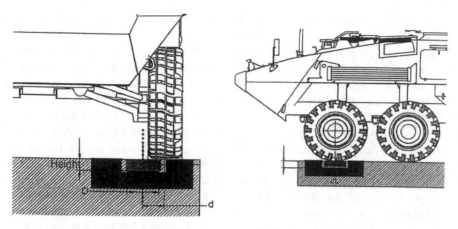

Fig. 1. The location of the explosive under the vehicle's wheel in a steel base, where: S- wheel's width, d- distance between the wheel's symmetry axis and the explosive, D- explosive's diameter [2]

and, by hitting the head against the vehicle's sides and roof, in spine injuries of the crew members. The head, face and neck areas constitute only 12% of the total body surface area which is exposed to injuries during fight. However, the injuries of that part of the body lead to serious disfunctions or death [5–7].

The injuries of lower limbs are connected with the effects of the shock wave on the frame construction. The shock wave, which affects the body, transfers energy between tissues of different acoustic impedance. This leads to cell disruption, soft tissue destruction and micro-fractures of bones. The example of the effects of the IED explosion on a soldier has been presented in Fig. 2.

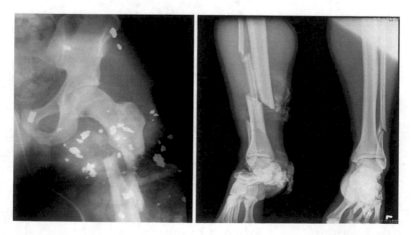

Fig. 2. The injuries of lower limbs resulting from the explosion of IED under the armoured vehicle [8].

182   G. Sławiński et al.

The tests regarding the resistance of the human body to the pulse load from a mine explosion should consist, first of all, of experimental tests and the identification of the effects of the blast wave on particular body areas. Due to limited information concerning the overloads coming from the effects of explosive detonations on humans and the inefficiently detailed replacement anatomical models of the areas of the human body or the use of replacement animal models with a significantly different body structure and different biochemical and metabolic mechanisms and the neuronal reaction, the key solution is to supplement the experimental results with computer numerical simulations.

Guidelines for conducting identification tests of overloads affecting the crew of military vehicles have been regulated in the HFM – 090/TG-25 document [9]. Apart from the methodology of conducting those tests, the document includes also the list of necessary equipment and the description of the ways of verifying the injuries of the crew caused by the explosion of land mines. The standard information regarding the mechanisms of injuries and the criteria of threats to crew's life concern: lower limbs, the Dynamic Response Index (DRI) of the thoracic and lumbar spine, the cervical spine and the head, the chest and internal organs. As a result of the detonation of an explosive under the vehicle, a soldier is exposed to injuries in many body areas, caused by the effects of sudden forces, moments and accelerations.

The experimental tests of the vehicle/dummy schemes exposed to the shock wave from the explosion of the AT mine require the use of at least one 50-centile anthropomorphic dummy (Hybrid III ATD) and sensors located in four critical places specified in Table 2. The critical values have been listed in Table 3. And the location of local coordinate systems of the dummy in a sitting position has been depicted in Fig. 3.

Table 2. Values registered by the sensors installed in the Hybrid III ATD dummy [2]

Sensor location	Measured value	Parameter description
Upper part of the neck	F_x	Horizontal force
	F_z	Vertical force
	M_y	Bending moment
Pelvis	A_z	Vertical acceleration
Lower part of shinbones	F_z	Vertical force
Chest	P_x	Overpressure

Table 3. The criteria of injuries and critical values [2]

Criterion no.	Body part	Criterion description	Parameter	Critical value
1	Lower limb	The maximum value of the axial compression force in the lower part of the shinbone	$(-F_z)$max	5.4 kN
2	Thoracic and lumbar	Dynamic Response Index calculated on the basis of the Az	DRIz	17.7

(continued)

Table 3. (*continued*)

Criterion no.	Body part	Criterion description	Parameter	Critical value
	part of the spine	horizontal acceleration of the pelvis		
3	Upper part of the neck	Axial compression force in the upper part of the neck considering the duration time	$[-F_z(t)]$max	4.0 Kn for 0 ms 1.1 kN for ≥ 30 ms
4a	Upper part of the neck	The moment of the upper part of the neck (bending forwards)	(M_{yp})max	190 Nm
4b	Upper part of the neck	The moment of the upper part of the neck (bending backwards)	$(-M_{yp})$max	57 Nm
5	Internal organs	Chest bending speed index	CWVP	3.6 m/s

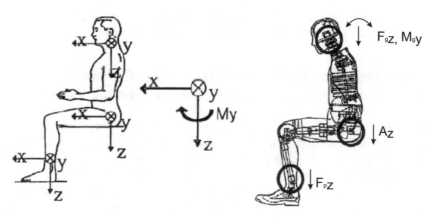

Fig. 3. Hybrid III in a sitting position with the defined coordinate system and marked measured values [2]

4 Injuries of Lower Limbs

The probability of the fracture of the shinbone has been estimated through the experimental tests of the bones of lower limbs [9, 10]. That relation has been presented in Fig. 4 in the age function and the values of the dynamic axial load measured in the distal direction of the bones. The fracture of the shinbone has been assumed in the developed analytical and physical models as the most reliable assessment criterion of the injuries of lower limbs. The shinbone is the basic bone of lower legs and transfers the biggest shock loads. The injury-fracture of the shinbone results in the lack of possibility for a given person to move independently and involves long rehabilitation.

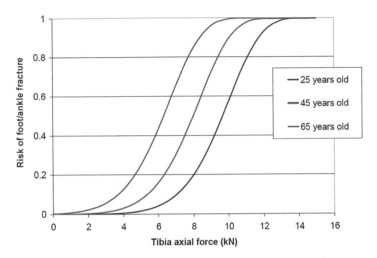

Fig. 4. The diagram presenting the probability of injury in the load function (Based on Data from Yoganandan 1996) [9]

In addition, it is possible to easily conduct experimental verification, which in the case of metatarsal bones or the knee or ankle joint, is a difficult task.

5 Tests Regarding the Effects of the Wave of Pressure from an Explosion on the Driver of the Military Vehicle

The aim of using the FEM numerical methods was to render it possible to correctly test the phenomenon of an explosion in its effects not only on the vehicle's supporting structure but also, indirectly, on the driver. Thanks to the conducted numerical analyses, it was possible to learn about the effects of an explosion on a soldier and assess the risk of injuries in relation to the increase of the seize of the used explosive.

The numerical model of the hull of the light armoured vehicle (with the driver) has been selected as the basis for numerical analyses. It has been subjected to the effects of the shock wave coming from the detonation of TNT explosives with different mass. The numerical simulations have been conducted using the LS-Dyna calculation code [11]. The numerical tests have been limited to multi-variant numerical calculations, in which the simplified model of the light armoured vehicle has been used (Fig. 5).

6 Numerical Model of the Military Vehicle with the Driver

The model of the simplified light armoured vehicle has been prepared using the following software: CATIA V5, HyperMesh, LS_PrePost. The LS_Dyna software has been used for the numerical calculations. The detonation and loading with the shock wave have been conducted by using the built-in CONWEP option [11].

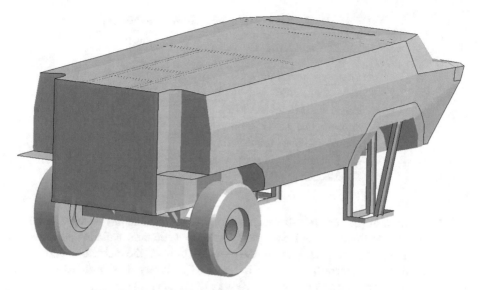

Fig. 5. The simplified model of the light armoured vehicle

The ELFORM16 – "Full integrate point" surface elements, used for the plating of the vehicle, have been used for the discretisation of the model. Other parts of the vehicle have been divided into finite elements by using the solid elements with one ELFORM1 integration point.

The loading with an explosion has been conducted by using the explosive function of CONWEP. In order to do that, it is necessary to declare a surface which the shock wave will have an impact on. The surface is declared by a segment whose normal vectors should be pointed to the explosive. Welded connections have been replaced by the TIDE contact connection, which prevents the displacement and turning of the selected nodes in relation to the selected surface. Other elements have been assigned the options of the AUTOMATIC_SINGLE_SURFACE contact. All of the mentioned contact options use the procedure based on the "penalty function" method.

The used material constants (units: mm, s, t, N, MPa, mJ) and other options have been presented in Tables 4 and 5 .

Table 4. Material constants for steel S355 [12]

Parameter	Steel
Mass density, RO	7.8e–9
Young's modulus, E	2.1e5
Yield stress, SIGY	355
Plastic strain to failure, FAIL	0.6
First effective plastic strain value EPS1	0.0
Second effective plastic strain value	0.6
Corresponding yield stress value to EPS1	355
Corresponding yield stress value to EPS2	550

Table 5. Material constants for steel ARMSTAL 500 [12]

Parameter	Steel
Mass density, RO	7.8e–9
Young's modulus, E	2.1e5
Yield stress, SIGY	1350
Plastic strain to failure, FAIL	0.5
First effective plastic strain value EPS1	0.0
Second effective plastic strain value	0.6
Corresponding yield stress value to EPS1	1350
Corresponding yield stress value to EPS2	1500

The position of the vehicle's driver in relation to the vehicle has been presented in Figs. 6 and 7. The driver's feet have been placed on a foothold which is supposed to simulate a floating protective part of the construction of the vehicle's body. The Hybrid III dummy, generally available in the LSTC library, has been used for the numerical calculations. The mass of the Hybrid III dummy was 75 kg.

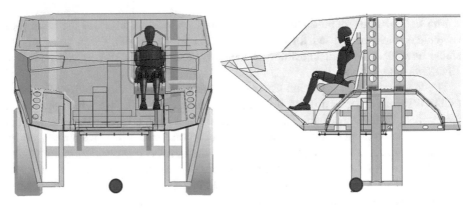

Fig. 6. The position of H3 and the explosive in relation to the replacement military vehicle

7 Validation of the Numerical Model

In order to authenticate the results of the numerical analyses, the validation of the model has been conducted, using the results of the field tests in which the body of the military vehicle has been loaded with the blast wave coming from the detonation of the Semtex 1A explosive with the mass of 0.5 kg. The explosive has been placed under the vehicle in a way corresponding to the experimental tests (Fig. 8).

For the purposes of the validation of the numerical models, the results of the experimental tests conducted with the participation of the Military Institute of Armoured and Automotive Technology have been used. Those results are presented in Figs. 9 and 11. In order to do so, the characteristics of the changes of the compression force in the right and left lower limb have been compared. The results obtained through

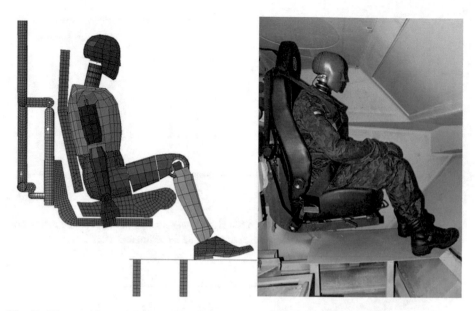

Fig. 7. The position of Hybrid III 50th% in the KORPUS vehicle and the corresponding position of the Dummy 50th%

Fig. 8. The view of the vehicle with the location of the explosive

the FEM analyses, corresponding to the conducted experimental tests, have been depicted in Figs. 10 and 12. The maximum values of the compared values have been presented in Tables 6 and 7.

The value of the compression force registered in the right lower limb of the HYBRID III anthropomorphic dummy during the experimental tests was 404 N, while the numerical analysis resulted in the analogous value amounting to 645 N (Table 6), which constitutes a relative error at the level of 38%. In the case of the left lower limb, those values were respectively 762 N (experimental tests) and 595 N (results from the

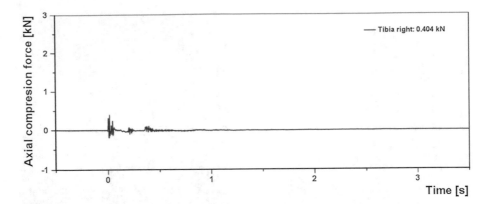

Fig. 9. The value of the compression force in the right lower limb in the case of the detonation of a 0.5 kg explosive under the vehicle – the results of the experimental tests

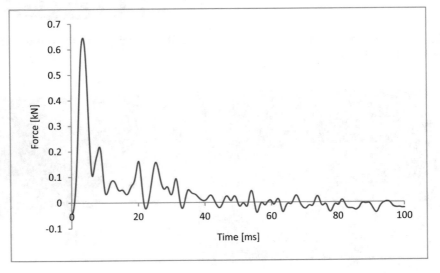

Fig. 10. The value of the compression force in the right lower limb in the case of the detonation of a 0.5 kg explosive under the vehicle – the result of the numerical calculations

numerical simulations). In that case, a relative error was at the level of 22% (Table 7). Taking into consideration the complexity level of the model as well as the significant simplifications of the validation, the result of the validation should be considered satisfactory.

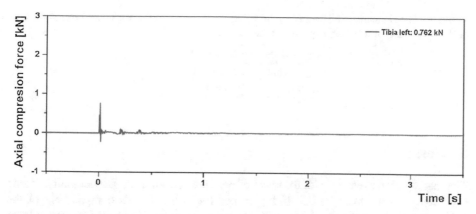

Fig. 11. The value of the compression force in the left lower limb in the case of the detonation of a 0.5 kg explosive under the vehicle – the results of the experimental tests

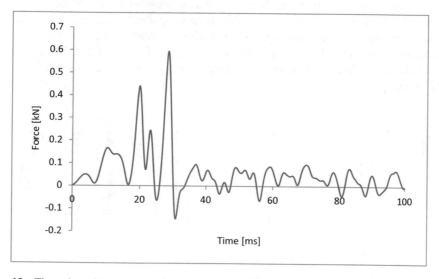

Fig. 12. The value of the compression force in the left lower limb in the case of the detonation of a 0.5 kg explosive under the vehicle – the result of the numerical calculations

Table 6. The comparison of the maximum values of the force in the right lower limb in the case of the detonation of a 0.5 kg explosive under the vehicle

Value of the compression force	
Experiment	0.404 kN
MES	0.645 kN

Table 7. The comparison of the maximum values of the force in the left lower limb in the case of the detonation of a 0.5 kg explosive under the vehicle

Value of the compression force	
Experiment	0.762 kN
MES	0.595 kN

8   Results

The loading of the vehicle with the wave of pressure coming from the detonation of the explosive with the mass of 0.5–10 kg located under the drive's compartment of the vehicle's body has been simulated in the numerical analyses. The load has been created using the CONWEP algorithm. The distance of the explosive from the floor surface was 700 mm.

The conducted numerical analyses rendered it possible to assess the risk of the injuries of lower limbs of the driver of the military vehicle loaded with the impulse of pressure generated by the detonation of the explosive under the vehicle. The results of the numerical analyses, for the validated numerical model, in the form of the courses of the force changes in the shinbone and the thigh have been presented in Figs. 13 and 14 and in Table 8.

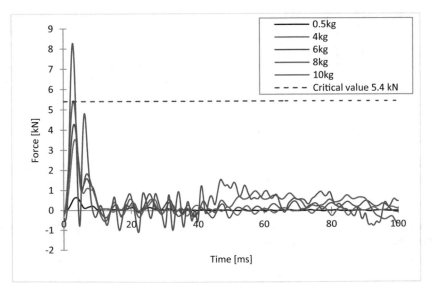

Fig. 13. The value of the force in the right lower limb – the results of the numerical calculations

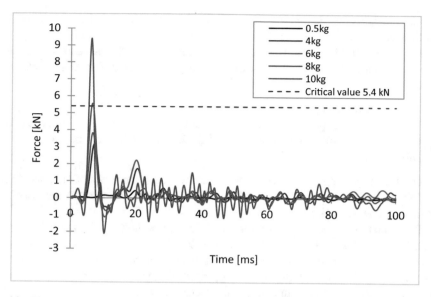

Fig. 14. The value of the force in the left lower limb – the results of the numerical calculations

Table 8. Maximum forces in the lower limbs for different loads of explosives

Mass of explosive charge	0.5 kg	4 kg	6 kg	8 kg	10 kg
Right lower limb	0.64	3.53	4.28	5.43	8.27
Left lower limb	0.59	3.17	3.82	5.56	9.40

It may be observed, on the basis of the achieved characteristics, that the maximum value of the compression force in the shinbone increases together with the increase of the explosive's mass, which is an obvious phenomenon. In the case of the effects of the shock wave created as a result of the detonation of explosives with the mass of 0.5–6 kg, the registered values both in the left and right lower limb have not exceeded the injury criterion for that part of the body (5.4 kN), whereas in the case of the other two variants (6 and 8 kg), that value has been exceeded. This would result in the injury of the lower limb in the form of contusions, fractures and tendon and ligament ruptures, which require medical intervention. Therefore, the key task of a designer of the vehicle is to develop appropriate solutions to level dangerous consequences of an attack on occupants inside the vehicle.

9   Summary

Based on the achieved results and the information included in the NATO normative documents regarding the level of protection of military vehicles against mines, it may be concluded that the provision of appropriate protection is a result of multiple factors, i.e. the vehicle's construction, the ergonomics of the interior and the equipment which would play the role of the absorbent of the energy coming from an explosion.

The values of the forces in lower limbs achieved from the numerical calculations are very similar to the values from the experimental tests. The differences in results may be caused by:

- the simplifications of the numerical model
- differences in the positioning of the Hybrid 3 dummy. Small differences in the positioning of lower limbs have resulted in different values of the forces in lower limbs.

The tests regarding the effects of the shock wave on the body of occupants inside the vehicle require a broad view of the sources of loads which are the reasons of injuries (in that case, the injuries of lower limbs).

The reduction of the interaction force should consist, first of all, in developing such a construction of the vehicle which minimises the danger of damaging the plating and which renders it possible to maximally dissipate the shock wave by the vehicle's construction or the use of additional energy-absorbing panels. In practice, the fact that the vehicle's hull is not damaged results in the increase of survivability of those injured, however, as a result of the displacement of the vehicle and the deformation of the hull, in particular the floor, and thus transferring loads to the limbs, it is still a significant problem.

Acknowledgements. The research was done within project no. DOBR-BIO4/022/13149/2013 'Improving the Safety and Protection of Soldiers on Missions Through Research and Development in Military Medical and Technical Areas', supported and co-financed by NCR&D, Poland.

References

1. Krzystala, E., Mężyk, A., Kciuk, S.: Analiza zagrożenia załogi w wyniku wybuchu ładunku pod kołowym pojazdem opancerzonym. Zeszyty naukowe WSOWL Nr 1(159) 2011
2. AEP-55: Procedures for Evaluating the Protection Levels of Logistic and Light Armoured Vehicles for KE and Artillery Threats, vol. 1, 1st edn. NATO/PFP Unclassified (2005)
3. AEP-55: Procedures for Evaluating the Protection Levels of Logistic and Light Armoured Vehicle Occupants for Grenade and Blast Mine Threats Level, vol. 2, 1st edn. NATO/PFP Unclassified (2005)
4. DGLEPM T & E Engineering Std: Improvised Explosive Device Protection Systems, LOI/P&A for TAPV Project, Unclassified (2010)
5. Wade, A.L., Dye, J.L., Mohre, Ch.R., Galarneau, M.R.: Head, Face, and Neck Injuries During Operation Iraqi Freedom II: Results From the US Navy and Marine Corps Combat Trauma Registry. San Diego, California. Technical Report 06-01
6. Sławiński, G., Niezgoda, T., Barnat, W., Wojtkowski, M.: Numerical analysis of the influence of blast wave on human body. J. KONES **20**, 381–386 (2013)
7. Mackiewicz, A., Sławiński, G., Niezgoda, T., Będziński, R.: Numerical analysis of the risk of neck injuries caused by IED explosion under the vehicle in military environments. Acta Mech. Autom. **10**, 258–264 (2016)

8. Wojtkowski, M., Ziółek, J., Płomiński, J., Waliński, T.: The army security in the aspect of threats resulting from the use of improvised explosive devices, red. In: Kowalkowski, S., Bębenek, B., Całkowski, T. (ed.) Postexplosion Analysis Musculoskeletal Injuries of Polish Soldiers in Military Contingents – Preliminary. National Defence University Publishing House (2014)
9. RTO-TR-HFM-090 AC/323(HFM-090)TP/72: Test Methodology for Protection of Vehicle Occupants against Anti-Vehicular Landmine Effects, NATO Research and Technology Organization (2007)
10. Tajszerska, D., Świtoński, E., Gzik, M.: Biomechanika narządu ruchu człowieka, Wydawnictwo Katedra Mechaniki Stosowanej, Politechnika Śląska, Gliwice 2011 LS-Dyna V971. Livermore Software Technology Corporation, Livermore (2006)
11. Świerczewski, M., Sławinski, G.: Modelling and numerical analysis of explosion under the wheel of light armoured military vehicle. Eng. Trans. 65(4), 587–599 (2017)
12. Slawinski, G., Swierczewski, M., Malesa, P.: Modelling and numerical analysis of explosion underneath the vehicle. J. KONES 24(4), 279–286 (2017)

Modeling Viscoelastic Behavior of Pig's Skin in the Respect to Its Anisotropy

Aneta Liber-Kneć[(✉)] and Sylwia Lagan

Cracow University of Technology, Warszawska 24, 31-155 Cracow, Poland
aliber@pk.edu.pl
http://pk.edu.pl

Abstract. The purpose of this work was to analyze the influence of different strain levels on the results of fitting the quasi-linear viscoelastic material model to the experimental curves. The stress relaxation test of pig's skin tissue was realized. Three values of strain levels were set at 5, 10, 15%. The anisotropy of material was taken into account. Correlation coefficients of fitting were evaluated. The study confirmed both the relationship between the level of strain, orientation of specimens and the values of model parameters.

Keywords: Stress relaxation · QLV model
Orientation of specimen taken · Soft tissue

1 Intoduction

Skin is a complex tissue composed of several layers and is characterized by anisotropy, viscoelasticity and non-linearity [2,3]. Anisotropy influence on mechanical properties and extensibility of skin which must be considered during wounds closure. In most regions of the body, there is skin tension in every direction, but the degree of tension is greatest parallel to the relaxed skin tension lines (RSTLs). The lines of maximal extensibility run perpendicular to RSTLs and represent the direction in which closure can be performed with the least tension. Viscoelastic properties allow the skin to adapt to forces applied to it. The stress relaxation and creep are two main mechanisms which result from this adaptability and may assist in wound closure. Stress relaxation refers to a decrease in stress that occurs when skin is held under tension at a constant strain. If a skin flap is closed under excessive tension, a certain amount of relaxation occurs as the tissue creeps [1,4,5]. This study aims to test the applicability of the quasi-linear viscoelastic (QLV) model to describe stress relaxation of skin considering its anisotropy. To identify viscoelastic behavior of skin tissue the QLV model introduced by Fung is commonly used [6]. It the skin collagen network of crimped fibers rotate to align the axis of loading. This has a different influence on skin behavior and may affect efficacy of the QLV theory to model stress relaxation of skin [7].

© Springer Nature Switzerland AG 2019
K. Arkusz et al. (Eds.): BIOMECHANICS 2018, AISC 831, pp. 194–201, 2019.
https://doi.org/10.1007/978-3-319-97286-2_17

2   Material and Methodology

2.1   Sample Preparation

In the study, the skin from a domestic pig, which weighed ca. 165 kg and was 9 months old, was used. Firstly, patches of skin from the back were extracted. The skin and adipose tissues were separated. Rectangular samples in two orientations with respect to the long axis of the pig's body: parallel and perpendicular were cut. All samples had the same geometric dimensions: 100.0 ± 0.2 mm in length and 10.0 ± 0.2 mm in width. The average thickness was equal to 2.2 ± 0.2 mm. In order to better store the biological material, the specimens were frozen at the temperature of $-18\,°C$ and defrosted in the period of 1 h at room temperature before the test [8].

2.2   Stress Relaxation Test

The stress relaxation test was determined with the use of the MTS Insight 50 testing machine. The samples were mounted using scissor action grips with self-tightening. The measurement base of the sample was 50 mm. Each set of samples for stress relaxation testing (the samples divided according to the orientation of their taking) contained a minimum of 3 samples. The preload of 1 N was applied to skin samples in order to get close to its natural state of pretension. Constant strain of 5, 10 and 15% was applied respectively for different samples. Stress relaxation behavior was observed until the value of stress decreased below the value of pre-stress (0.05 MPa). The stress relaxation curves were shown as a dependence of true stress and real time or normalized stress and real time.

2.3   Quasi-Linear Viscoelestic (QLV) Characteristic [6, 7]

In this study, relaxation curves were shown using the reduced relaxation function of the sample, defined as (1) and (2):

$$G(t) = \frac{\sigma(t)}{\sigma_{max}} \tag{1}$$

$$G(0) = 1 \tag{2}$$

where $\sigma(t)$ is the stress at time t, σ_{max} is the initial stress, amplitude at $t = t_p$ corresponding to the maximum stress. In the QLV constitutive model, the instantaneous stress is given by the Eq. (3):

$$\sigma(t) = G(t) \cdot \sigma^e(\varepsilon) \tag{3}$$

where $\sigma^e(\varepsilon)$ is the stress at temporary strain. The time-dependent mechanical behavior of soft tissues can be written as (4):

$$\sigma(t) = \int_0^t G(t - \tau) \frac{\partial \sigma^e(t)}{\partial \varepsilon} \frac{\partial \varepsilon}{\partial \tau} \tag{4}$$

where $\frac{\partial \sigma^e(t)}{\partial \varepsilon}$ is the temporary elastic response, and $\frac{\partial \varepsilon}{\partial \tau}$ is the strain history of the sample. A series of exponentials, also known as the Prony series, is a commonly used relaxation function. The reduced relaxation function is given as (5):

$$G(t) = A_0 + A_1 e^{-t/\tau_1} + A_2 e^{-t/\tau_2} + A_3 e^{-t/\tau_3} \tag{5}$$

where A_0, A_1, A_2, A_3, τ_1, τ_2, τ_3 are constants, which can be determined by using experimental data. An exponential (AB model) non-linear elastic representation of the stress response is given as (6):

$$\sigma^e(\varepsilon) = A(e^{B\varepsilon} - 1) \tag{6}$$

where A is a linear parameter which has the same dimension as stress, and B is non-dimensional factor describing the nonlinearity of elastic response. The optimization procedure was performed with the use of the OriginLab to generate data fitting curves for samples of skin tissue. The QLV model was used to fit the stress-time data both the ramp and relaxation phase.

3   Results and Discussion

The experimental stress relaxation curves (with average curve) for three levels of strain were shown in Fig. 1 for samples taken perpendicular to back bone and for parallel samples in Fig. 2. A strong stress relaxation was observed at the beginning of the relaxation process for all specimens but the most significance decrease of stress occurred at 15% strain. The different orientations of samples taken influenced on stress relaxation, especially in the case of a higher level of strain (10 and 15%). The value of peak stress on imposition of ramp stimulus is higher for perpendicular orientation. The perpendicular samples achieved the following average values of initial stress: at 5% strain – 0.15 MPa for 30 s, at 10% strain – 0.67 MPa for 60 s, and at 15% strain – 1.16 MPa for 90 s. For parallel samples, the value of initial stress is 0.15 MPa for 30 s at 5% strain, 0.21 MPa for 60 s at 10% strain and 0.81 MPa for 90 s at 15% strain. The QLV model requires a reduced relaxation function thus the original relaxation function was first converted to normalized stress (1) and plotted against time. The QLV model fit the data well for all strains and both parallel and perpendicular orientations of samples with an average $R^2 = 0.99$ as seen in Fig. 3. Also the QLV model predictions of stress–time curves in ramp phase agreed well with the experimental data (Fig. 4). Results of calculation of the coefficients were shown in Table 1 for perpendicular orientation of sample taking and in Table 2 for parallel one. The value of the elastic component A increased with the strain, more for perpendicular samples than for parallel ones. The value of elastic component B slightly decreased with the strain for perpendicular samples and in lesser range for parallel samples. Good ability of the QLV model to predict pig's skin stress relaxation process for the different orientations of samples cutting was observed by Liu et al. [7] however the values of calculated coefficient differ from these obtained in this study. It may result from pig's tissue diversity and preconditioning procedure.

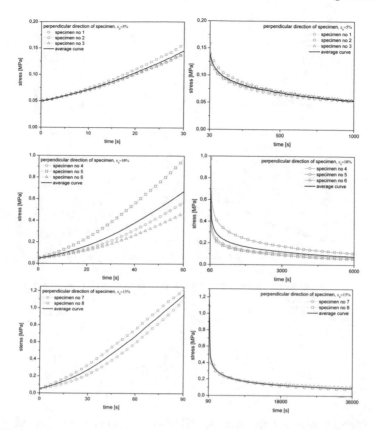

Fig. 1. The experimental data curves for all levels of strain and perpendicular orientation of specimens taking (left column – ramp, right column – stress relaxation curve)

Table 1. Parameters of models for perpendicular specimen orientation (mean $\pm$ SD)

Strain level	5%	10%	15%
A	0.057 $\pm$ 0.002	0.218 $\pm$ 0.136	0.612 $\pm$ 0.341
B	0.219 $\pm$ 0.010	0.156 $\pm$ 0.024	0.086 $\pm$ 0.003
A_0	0.244 $\pm$ 0.022	0.066 $\pm$ 0.010	0.070 $\pm$ 0.029
A_1	0.148 $\pm$ 0.030	0.400 $\pm$ 0.029	0.575 $\pm$ 0.025
A_2	0.222 $\pm$ 0.026	0.256 $\pm$ 0.022	0.205 $\pm$ 1E$-$05
A_3	0.385 $\pm$ 0.031	0.277 $\pm$ 0.024	0.151 $\pm$ 0.005
τ_1	10.472 $\pm$ 3.880	28.879 $\pm$ 2.159	129.355 $\pm$ 41.643
τ_2	97.799 $\pm$ 32.206	300.219 $\pm$ 33.209	1742.850 $\pm$ 742.914
τ_3	899.550 $\pm$ 267.871	3368.018 $\pm$ 690.014	14074.977 $\pm$ 4741.436

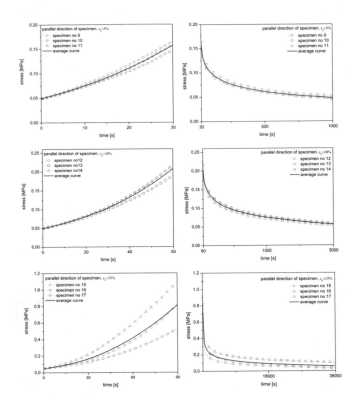

Fig. 2. The experimental data curves for all levels of strain and parallel orientation of specimens taking (left column – ramp, right column – stress relaxation curve)

Table 2. Parameters of models for parallel specimen orientation (mean ± SD)

Strain level	5%	10%	15%
A	0.070 ± 0.013	0.058 ± 0.002	0.093 ± 0.012
B	0.207 ± 0.028	0.149 ± 0.009	0.154 ± 0.021
A_0	0.256 ± 0.031	0.182 ± 0.009	0.047 ± 0.041
A_1	0.151 ± 0.037	0.181 ± 0.019	0.557 ± 0.028
A_2	0.250 ± 0.014	0.259 ± 0.009	0.213 ± 0.023
A_3	0.343 ± 0.019	0.378 ± 0.019	0.183 ± 0.022
τ_1	8.041 ± 0.976	22.054 ± 2.476	151.320 ± 45.423
τ_2	71.086 ± 11.425	212.332 ± 11.018	1726.986 ± 526.168
τ_3	515.740 ± 76.258	2024.905 ± 102.744	17033.122 ± 5507.822

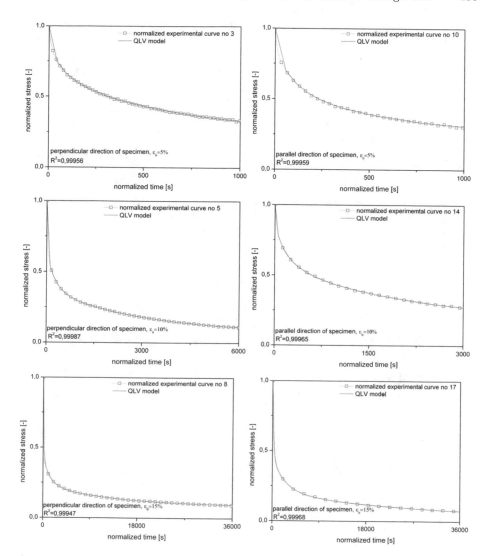

Fig. 3. Comparison of representative normalized curves obtained from experimental data with curves predicted through the QLV modelling

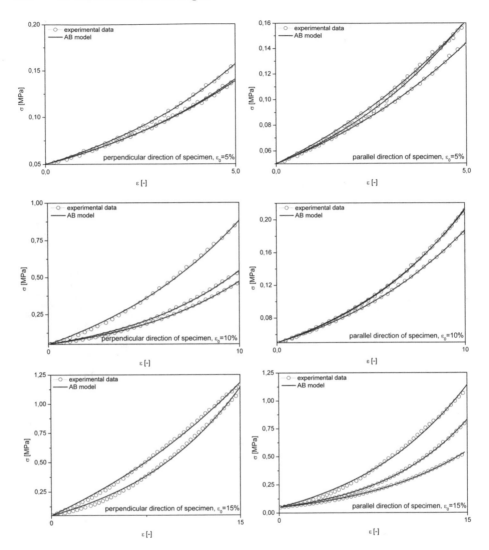

Fig. 4. Comparison of normalized curves obtained from experimental data with AB model fitting curves for ramp phase

4   Conclusions

Conducted tests and calculations demonstrated the influence of skin anisotropy both on stress relaxation process for pig's skin and values of material model parameters. Results of modeling with the use of the quasi-linear viscoelastic theory with the reduced relaxation function showed its ability to accurately capture experimental data of skin tissue. Due to the specificity of the research material which is soft tissue, future research should take into account the effect of preconditioning, other levels of deformation and different anatomical locations on the viscoelastic response. Such investigations in porcine models deliver the primary information that can be use in wound closing without tension, analyses of pathological changes in skin or thermal skin sensation.

Acknowledgement. The work was realized due to statutory activities M-1/12/ 2018/DS.

References

1. Baker, S.R.: Local Flaps in Facial Reconstruction, pp. 31–32. Elsevier, Philadelphia (2014)
2. Ni Annaidh, A.N., Bruyere, K., Destrade, M., Gilchrist, M.D., Maurini, C., Ottenio, M., Saccomandi, G.: Automated estimation of collagen fiber dispersion in the dermis and its contribution to the anisotropic behavior of skin. Ann. Biomed. Eng. **40**(8), 1666–1678 (2012)
3. Lagan, S., Liber-Kneć, A.: Experimental testing and constitutive modeling of the mechanical properties of the swine skin tissue. Acta Bioeng. Biomech. **19**(2), 93–102 (2017)
4. Bailey, B.J., Johnson, J.T., Newlands, S.D.: Head and neck surgery – otolaryngology, 2361 p. Lippincott Wiliam & Wilkins, Philadelphia (2006)
5. Topaz, M., Carmel, N.N., Topa, Z.G., Li, M., Li, Y.Z.: Stres-relaxation and tension relief system for immediate primary closure of large and huge soft tissue defects: an old-new concept. Medicine (Baltimore) **93**(28), 1–6 (2014)
6. Fung, Y.C.: Biomechanics: Mechanical Properties of Living Tissues. Springer-Verlag, New York (1993)
7. Liu, Z., Yeung, K.: The preconditioning and stress relaxation of skin tissue. J. Biomed. Parmaceutical Eng. **2**(1), 22–28 (2008)
8. Liber-Kneć, A., Lagan, S.: Testing stress relaxation process of a porcine skin. Eng. Biomate. **19**(134), 18–24 (2016)

FEM Analysis of Hyperelastic Behavior of Pig's Skin with Anatomical Site Consideration

Sylwia Łagan$^{(\boxtimes)}$, Agnieszka Chojnacka-Brożek, and Aneta Liber-Kneć

Cracow University of Technology, Warszawska 24, 31-155 Cracow, Poland
slagan@mech.pk.edu.pl,
http://pk.edu.pl

Abstract. The aim of this work was to compare a hyperelastic material models (Ogden and Yeoh) implemented in ANSYS software to evaluate the experimental data obtained from uniaxial tensile test of pig's skin taken from different location of the body (back and abdomen). The results were compared with the literature data. The effects of the simulation confirmed the possibility of using hyperelastic models to assess skin properties using FEM methods based on a static tensile test. The study reveals the need to validate results based on a wider range of data.

Keywords: Tensile test · Non-linear static analysis · FEM
Correlation of mechanical properties · Soft tissue

1 Introduction

To identify the mechanical reaction occuring in biological materials during loading, a pig's skin is often used as the substitute of human tissue. The pig's tissue is used in graft and, after process of acellularization, is used as scaffolds in the tissue engineering. The study of the mechanical properties of skin (ultimate tensile stress, Young modulus, extensibility or stress relaxations) as well as accurate predictions of its mechanical behavior are desired in many surgical processes [5,12]. The corellation of calculated mechanical properties with the actual behavior is crucial for prediction of short and long–term mechanical behavior [6,8]. The nonlinearity and anisotropy of skin tissue have significant influence on its mechanical behavior. Additionally, it strongly depends on the location on the body and on the individual [2]. The FEM (finite element method) is useful to visualize a distribution of stress, strain or displacement areas under mechanical loading fields [11]. However, there is not a widely used approach in modelling skin tissue [1].

In this work, the FEM method was used to estimate two hyperelastic multi–parameters material models: the Ogden and the Yeoh. Both models are widely used to determine relations of hyperelastic materials such soft biological tissues as brain, lung, liver and spleen [12]. The applicability of modelling procedure was realized with the use of tensile test for pig's skin presented in [8]. The anatomical location of pig's skin taken was considered in the study, too.

© Springer Nature Switzerland AG 2019
K. Arkusz et al. (Eds.): BIOMECHANICS 2018, AISC 831, pp. 202–209, 2019.
https://doi.org/10.1007/978-3-319-97286-2_18

2   Material and Methodology

2.1   Test Data

The sample preparation and procedure of uniaxial tensile test as well as the procedure of the strength parameters evaluation were the object of earlier researches, presented in [8]. In the FEM simulations, experimental averages stress–strain curves were used.

2.2   Hyperelastic Material Model Characteristic

In this study, in order to describe the nonlinear tensile stress–strain behavior of skin tissue the Ogden model and the Yeoh model were used. Generally a hyperelastic material model is based on the strain energy function. Depending on the class and type of material, strain energy function can be expressed in different ways. The material can be regarded as isotropic, incompressible and hyperelastic. The strain–energy deformation *(W)*, assuming isotropy, can be described as depended on the strain invariants of the deformation tensor of Cauchy–Green I_1, I_2, I_3 defined as (1):

$$W_{iso} = W(I_1, I_2, I_3) \tag{1}$$

where:

$$I_1 = \lambda_1^2 + \lambda_2^2 + \lambda_3^2 \tag{2}$$
$$I_2 = \lambda_1^2\lambda_2^2 + \lambda_2^2\lambda_3^2 + \lambda_3^2\lambda_1^2 \tag{3}$$
$$I_3 = \lambda_1^2\lambda_2^2\lambda_3^2 \tag{4}$$

and $\lambda_1, \lambda_2, \lambda_3$ are the principial stretches.
Assuming the incompressible material and $\sigma_1 = \sigma_2$ for uniaxial tensile we get a simplified formula (2)–(4):

$$\lambda_1 = \lambda, \quad \lambda_2 = \lambda_3 = \frac{1}{\sqrt{\lambda}} \tag{5}$$

The left tensor invariant of deformation is simplified too, and expresses as:

$$I_1 = \lambda^2 + 2\frac{1}{\lambda}, \quad I_2 = \lambda^2 + 2\frac{1}{\lambda^2}, \quad I_3 = 1 \tag{6}$$

The strain–energy function in the Ogden material model depends on principial stretches [10], and is given as (7):

$$W = \sum_{m=1}^{N} \frac{2\mu_m}{\alpha_m^2}(\lambda_1^{\alpha_m} + \lambda_2^{\alpha_m} + \lambda_3^{\alpha_m} - 3) \tag{7}$$

where α_m and μ_m are the material constants to be determined.
In the Yeoh material model the strain–energy function is dependent on the first strain invariation, is viewed as a special case of the reduced polynomial with N = 3 and written as (8), where C_{i0} are the material constants.

$$W = \sum_{i=1}^{N} C_{i0}(I_1 - 3)^i \tag{8}$$

2.3   Conditions of FEM Simulations

The ANSYS software was used to evaluation of the value of hyperelastic material parameters. The test data, for uniaxial tensile tests, were specified and hyperelastic material models were evaluated to fit with experimental data for the calibration of six hyperelastic material models, the Ogden (1st, 2nd, 3th order) and the Yeoh (1st, 2nd, 3th order). To verify this, parameters for the uniaxial test carried out on the pig's skin strips were modelled by the finite element technique. The analyses were displacement controlled and the force required to achieve this extension was computed. The conditions of simulation were defined by the level of initial extension, respectively for back specimen models equal 4 mm, and for abdomen specimen equal 35 mm, applied on one side of the model and the other side was completely fixed. These values of extension were determined based on the results of the uniaxial extension test, as middle strain values of the range of the second stretching phase (nonlinear phase transition) [8]. The dimensions of specimen models were the same as experimental ones (abdomen: $1.5 \times 10.0 \times 50.0$ mm; back: $2.3 \times 10.0 \times 50.0$ mm). Before starting the hyperelastic modelling, convergent and accurate meshes have to be determined. Finally the geometries were discretized into 20–noded hexahedral elements, this type of element has greater performance than the tetrahedral elements especially for hybrid elements. The total number of elements was 2 856 with 15 189 nodes.

3   Results and Discussion

In Table 1, the results of the FEM modelling were compared with available literature data. The examples of stress and deformation distributions obtained by the use of the Ogden model on the surface for specimens respectively from the back were shown in Fig. 1, for abdomen specimen in Fig. 2 (the contours of initial dimensions of models were marked by black line).

Table 1. Parameters of material models for skin tissue of pig – back and abdomen location of body

Model	Parameter	Abdomen	Back	[8]
Ogden 1st Order	α	0.015	1.741	$0,057^{a}$, $3,580^{b}$
	μ	7.730	12.240	$7,728^{a}$, 27.413^{b}
Yeoh 3th Order	C_1	−0.015	1.620	-0.019^{a}, $0,687^{b}$
	C_2	0.059	87.258	$0,052^{a}$, 103.815^{b}
	C_3	0.0205	−184.65	0.003^{a}, -400.051^{b}

a–abdomen, b–back

The results of FEM analysis with application of the Yeoh model were presented for back site specimen in Fig. 3 and for abdomen specimen in Fig. 4. The presented figures showed differences well in behavior of samples from various

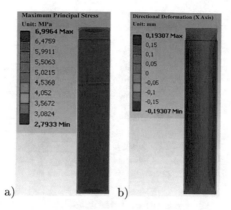

Fig. 1. The distribution of stress (a) and deformation (b) specimen (back) using the Ogden 1st order model

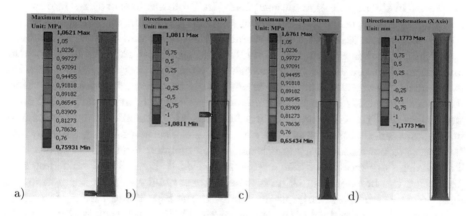

Fig. 2. The distribution of stress and deformation specimen (abdomen) using the Ogden respectively 1st order model (a,b) and 3th order (c,d)

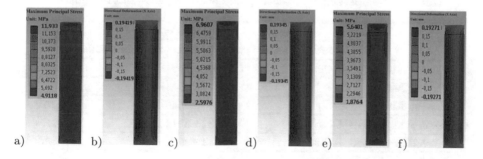

Fig. 3. The distribution of stress and deformation specimen (back) using the Yeoh model respectively 1st order model (a,b), the 2nd (c,d) and 3th order (e,f)

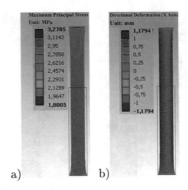

Fig. 4. The distribution of stress (a) and deformation (b) specimen (abdomen) using the Yeoh 1st order model

body site, in the middle of nonlinear phase. In Table 2 values of stress parameters obtained by fitting models implemented in ANSYS were shown. Additionally, the curves of maximal values of stress obtained in the FEM analysis for all calculated hyperelastic models were shown in Fig. 5. The non–linear curve fitting algorithm implemented in the Ansys code was used for this purpose. Figure 5 shows the experimental stress vs. strain curves with their corresponding predicted values from each model. As it can be seen models can closely fit the experimental data, the Ogden and Yeoh 2nd and 3th order models give reasonable degree of agreements between real and computed data but the degree of fitting for the Yeoh 1st model is not satisfactory and is not recommended for further calculations. The Ogden and Yeoh 3th order models appeared to be most suitable choice for predicting the behaviour of skin because of their ability to match experimental data points at small and large strain values. In the Figs. 6 and 7 fitting curves for experimental average data with the use of the Ogden and Yeoh material model were presented.

Table 2. Comparison of obtained values of stress [MPa] in FEM analysis

Model	Back			Abdomen		
	Min.	Specimen middle	Max.	Min.	Specimen middle	Max.
Ogden 1st	2,793	3,459	6,996	0,759	0,919	1,062
Ogden 2nd	–	–	–	–	–	–
Ogden 3th	–	–	–	0,654	0,795	1,676
Yeoh 1st	4,912	6,046	11,933	1,801	2,453	3,279
Yeoh 2nd	2,598	3,179	6,961	–	–	–
Yeoh 3th	1,876	2,311	5,640	–	–	–

The review of literature is rather poor. According to Flynn et al. [3] the Ogden model parameters were: $\mu = 0.0096$–0.0398 and $\alpha = 33.45$–35.99 for human skin

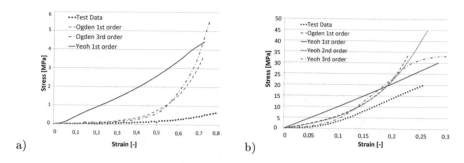

Fig. 5. Comparison fitting curve for applied models: (a) abdomen location, (b) back location

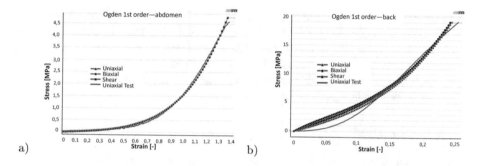

Fig. 6. Fitting curve for the Ogden model: (a) abdomen location, (b) back location

from posterior upper arm and anterior upper forearm. Remache (2018) reported $\mu = 0.02$ and $\alpha = 29$ for porcine abdomen skin. Chanda et al. (2015) referred results for the human skin surrogate and the Yeoh model and showed the values of $C_1 = 0.95$, $C_2 = 4.94$, $C_3 = 0.01$. The results of the FEM simulation showed some problems in compatibility of applied models to the experimental stress–strain curves. Some models did not achieve convergence of analysis results with experimental data. No stability was shown in cases of Ogden models (2nd order) for both anatomical locations (abdomen and back) and Ogden 3th for the back area, as well as using the Yeoh model (2nd and 3th order) for samples taken from the abdomen area. According to authors the time parameter should be consider as also the others shear and compression stresses. There are a number of limitations associated with this study [4]. Future work, will consider orientation and orthogonality of samples to the RSTL's (relaxed skin tension lines), in aim to improve the quality of samples used in experiment. The more experimental data obtained from different methods of characterization, such as tensile, compression, shear, torsion or suction tests will be considered in next studies. Also dynamical loading and stress relaxation tests should be taken into account. The strictly boundary conditions and the direction of loading and loading rate allows to proper identify the mechanical properties [6,7,9]. Finally, the sample form and size should be taken into account.

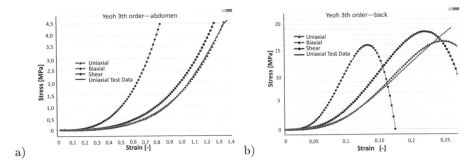

Fig. 7. Fitting curve for the Yeoh model: (a) abdomen location, (b) back location

Starting from in vitro experiments, through fitting constitutive hyperelastic material models and also increasing accuracy of FEM models, the global mechanical behavior of skin tissue under mechanical load will be evaluated. A next task is consideration of the effect of strain rate and orientation of the samples. The separation of the effect of the orientation and the effect of the strain rates in this loading region is therefore difficult.

4   Conclusions

The strength analysis by FEM has shown the higher values of maximum principal stress for specimen of pig's skin from back than abdomen via 1st order Yeoh model, the values respectively are 11.93 [MPa] and 3.28 [MPa]. Also the 1st order Ogden model has shown higher values for specimen of pig's skin from back, almost 7 [MPa] than abdomen specimens 1.06 [MPa], although higher (35 mm) initial extention for abdomen specimen. FEM analysis is a computational method to predict and to understand the mechanical behavior of skin. The numerical validation of the hyperelastic constitutive model was completed by means of comparisons between results from simulating experiment and experimental data. The comparisons showed reasonably good agreement for some of the models but also showed their limitations in comparison with experimental data. In view of the obtained results, it can be concluded that the Yeoh model was more accurate to reproduce the mechanical behavior of pig's skin specimens from spinal area, while the Ogden model of the abdomen skin specimens. The validity of the models in low or large strains will be the topic of the future studies. Improvement of efficient numerical procedures for predicting mechanical behavior in real time may contribute to the promotion of pre–operatory simulation of surgical procedures. Future work will focus on development and validation of FEM material model for the skin, including modelling of anisotropy. Also new experimental protocols and in vivo deformation measurements of skin using digital image correlation will be realized.

Acknowledgement. The work was realized due to statutory activities M-1/12/2018/DS.

References

1. Benitez, J.M., Montans, F.J.: The mechanical behavior of skin: Structures and models for the finite element analysis. Comput. Struct. **190**, 75–107 (2017)
2. Chanda, A., Graeter, R., Unnikrishnan, V.: Effect of blasts on subject-specific computational models of skin and bone sections at various locations on the human body. AIMS Mater. Sci. **2**(4), 425–447 (2015)
3. Flynn, C., Taberner, A., Nielsen, P.: Mechanical characterisation of in vivo human skin using a 3d force-sensitive micro-robot and finite element analysis. Biomech. Model. Mechanobiol. **10**(1), 27–38 (2011)
4. Ni Annaidh, A.N., Bruyere, K., Destrade, M., Gilchrist, M.D., Maurini, C., Ottenio, M., Saccomandi, G.: Automated estimation of collagen fiber dispersion in the dermis and its contribution to the anisotropic behavior of skin. Annal. Biomed. Eng. **40**(8), 1666–1678 (2012)
5. Lapeer, R.J., Gassona, D., Karri, V.: Simulating plastic surgery: from human skin tensile tests, through hyperelastic finite element models to real-time haptics. Progress Biophys. Mol. Biol. **103**(2–3), 208–216 (2010)
6. Liber-Kneć, A., Lagan, S.: Factors influencing on mechanical properties of porcine skin obtained in tensile test-preliminary studies. Adv. Intell. Syst. Comput. **623**, 255–262 (2018)
7. Łagan, S., Liber-Kneć, A.: Application of the ogden model to the tensile stress-strain behavior of the pig's. Adv. Intell. Syst. Comput. **526**, 145–152 (2017)
8. Łagan, S., Liber-Kneć, A.: Experimental testing and constitutive modeling of the mechanical properties of the swine skin tissue. Acta Bioeng. Biomech. **19**(2), 93–102 (2017)
9. Łagan, S., Liber-Kneć, A.: Influence of strain rates on the hyperelastic material models parameters of pig skin tissue. Adv. Intell. Syst. Comput. **623**, 279–287 (2018)
10. Martins, P., Jorge, R.N., Ferreira, A.: A comparative study of several material models for prediction of hyperelastic properties: application to silicone-rubber and soft tissues. Strain **42**, 135–147 (2006)
11. Remache, D., Caliez, M., Gratton, M., Dos Santos, S.: The effects of cyclic tensile and stress-relaxation tests on porcine skin. J. Mech. Behav. Biomed. Mater. **77**, 242–249 (2018)
12. Wex, C., Arndt, S., Stoll, A., Bruns, C., Kupriyanova, Y.: Isotropic incompressible hyperelastic models for modeling the mechanical behaviour of biological tissues: a review. BioMed. Eng/Biomedizinische Technik. **60**(6), 577–592 (2015)

Trabecular Bone Microstructural FEM Analysis for Out-Of Plane Resolution Change

Artur Cichański[✉] and Krzysztof Nowicki

Faculty of Mechanical Engineering,
University of Technology and Live Sciences in Bydgoszcz,
Al. prof. S. Kaliskiego 7, 85-796 Bydgoszcz, Poland
arci@utp.edu.pl

Abstract. The paper presents comparison of two methods of voxel defining for trabecular bone structure modelling. Regular cubic voxels were considered, the size of which changed uniformly in three mutually perpendicular directions depending on the adopted resolution. Also elongated rectangular voxels were proposed, characterized with fixed in-plane resolution and variable length along out-of plane direction. For both types of voxels a number of analyses were performed using finite element method with structures of varied BV/TV values. It was stated that voxel dimension change in out-of plane direction allows for decreasing of a number of scans required for correct reflect of the modelled structure stiffness for the needs of numerical analyses.

Keywords: Trabecular bone structure · Apparent stiffness · μFEM

1 Introduction

Not only does the content of its mineral fraction, but also its structure, decide about the stiffness of the tested trabecular bone [1]. Numerical analyses of such structures are performed with finite element method FEM use surface or voxel approach [2]. Computer simulations are conducted using hexahedral or tetrahedral elements [3] [4]. In case of voxel models the manner of defining and voxel size has significant impact on such structure stiffness [5]. The most common approach assumes that the voxel has cubic shape of dimensions resulting from the applied resolution CT scanners [2]. The finite element method of trabecular structures analyses are carried out with a resolution of approximately 30 μm even if the CT scan is carried out at a resolution of 6 μm [6]. During analyses, the stiffness decrease of the modelled structure along the voxel size increase can be noticed [7]. For relatively big voxels there is a loss of trabecular connectivity in the model. The loss of trabecular connectivity can be compensated by thickening of the remaining structure. The paper [8] presents such thickening with three-dimensional morphological dilatation.

© Springer Nature Switzerland AG 2019
K. Arkusz et al. (Eds.): BIOMECHANICS 2018, AISC 831, pp. 210–218, 2019.
https://doi.org/10.1007/978-3-319-97286-2_19

The paper undertakes an issue of the impact of voxel defining manner on the value of mechanical parameters, specified with finite element method. At the same time, the compatibility of the voxel creation algorithm with the idea of CT scanner operation was tried. Standard method, in which the voxel is of a cubic shape, was adopted as the base method. The method assuming application of elongated voxels was proposed as an alternative method. Both approaches were verified during computer analyses. The structure samples were reconstructed based on micro-tomograph scans. Resolution decrease was simulated by increasing characteristic voxel dimension specific for the discussed method. It was demonstrated how resolution change impacts the change of the volume and apparent module for both methods.

2 Materials and Methods

2.1 Materials

The tests were performed with trabecular structures cut from human femoral neck. The samples were of cylindrical shape of $\phi = 9.75$ mm diameter and 7.74 mm height. Manner of obtaining samples was described in the paper [1]. The samples were subject to examination on micro-tomograph CT80 with 36 μm resolution. 5 samples were selected for analyses out of a series of 17. Representatives were selected from the samples ordered in series in relation to BV/TV [9], so their BV/TV would correspond to minimum and maximum values and quartiles Q1, Q2, Q3 for the whole series. Structure indicators for the selected samples are presented in Table 1.

Table 1. Structure indicators of samples selected for analysis

Sample	BV/TV	Tb.Th, mm	Tb.N, 1/mm	Conn.D	DA
min BV/TV	0.1554	0.1608	0.9658	1.441	2.401
Q1 of BV/TV	0.1834	0.1565	1.1717	3.579	1.871
Q2 of BV/TV	0.2012	0.1853	1.0861	2.158	2.171
Q3 of BV/TV	0.2416	0.1792	1.3488	4.172	1.882
max BV/TV	0.2980	0.2345	1.2711	2.579	2.107

2.2 Methods of modelling

Computer models of the examined trabecular bone structures were created based on a set of images obtained from the micro-tomograph Fig. 1. Subsequent scans were set every 36μm apart. Two manners of structure modelling were applied for tests.

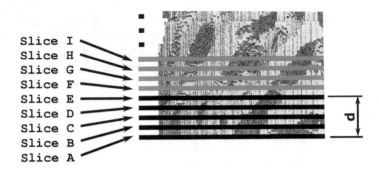

Fig. 1. Diagram of the set of micro-tomograph structure images

3D Method. In the first approach spatial resolution of the model changed uniformly, the voxel increased evenly in three mutually perpendicular directions. The approach, named as 3D method, assumes that the new voxel is created on the basis of subset of n-subsequent scans, positioned on d voxel length (e.g. scans A, B, C, D, E in Fig. 1). In such selected subset, the brightness of pixels, positioned in mutually corresponding positions along three mutually perpendicular directions, is aggregated. The calculated sums of brightness are scaled to the range of 0 to 1 with preservation of sensitivity threshold, assuming that the bone mineral fraction corresponding to value 1 is present when over half of pixels has been activated. This working algorithm, in the intention of the authors, well reflects the effects associated with the change in the CT scanner's working resolution. It also takes into account the effect of changing the amount of energy radiated by the camera. Obtained pixels are exchanged for voxels of a size resulting from multiplication of basic distance 36μm by number of n subset of subsequent scans, adopted for pixel definition. Exemplary slice of trabecular structure model for $n = 4$ times, corresponding to resolution 144 m is presented in Fig. 2. Finite element method mesh was created by direct conversion of geometry voxel to element [10]. In order to eliminate the impact of finite element size on calculation accuracy, characteristic size of the element in each analysis was fixed and amounted to 36μm. In the example presented in Fig. 2, voxel size changed in three directions, so each voxel was filled with $4^3 = 64$ finite elements.

FL Method. In the second approach in-plane resolution of the model change heterogeneously, voxel increased only in one out-of plane direction. This approach is called FL method and [5] and it assumes that the model in-plane resolution remained unchanged and corresponded to the initial scan resolution of 36μm. New voxel is created as a result of comparison of two scans located at distance d from each other, corresponding to voxel length (e.g. scans A and E in Fig. 1). If active pixels representing the bone mineral fraction are positioned on two compared layers, pixel in the resulting image is activated. Obtained pixels are exchanged to voxels of square basis, with 36 m side and the length corresponds to d dimension, describing distance between the compared scans. Exemplary slice

Fig. 2. A fragment of structure model reconstructed with 3D method, BV/TV = 0.155, d = 144 μm

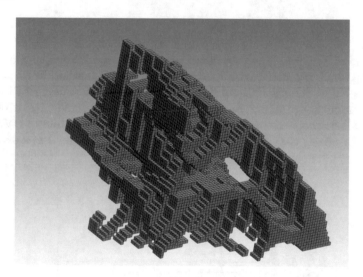

Fig. 3. A fragment of structure model reconstructed with FL method, BV/TV = 0.155, d = 144 μm

of a model with 144 μm resolution of trabecular structure characterizing with BV/TV = 0.155 is presented in Fig. 3. In this example the voxel size changed in one direction, so each voxel has been filled with $4^1 = 4$ finite elements.

2.3 FEM Analyses

Numerical analyses were performed in ANSYS v18.2 software environment. 8 node elements SOLID186 were used for analyses. Isotropic material properties $E = 10$ GPa and $\nu = 0.3$ were adopted for the analyses [11]. Boundary conditions mapped compression of cylindrical sample in axial direction $\epsilon = 0.8\%$ [5]. Apparent Young modules for the sample was specified based on the results of the analyses.

For each of the selected samples, presented in Table 1, structure models were generated for 3D and FL method. The models were subject to discretisation with elements of cubic shape with 36 μm side. The resulting mesh was deprived of the elements which lost connection with the whole sample structure. Numerical analyses were performed for voxels of length d changing from 72 to 288 μm with 36 μm increment. For each considered sample reference values were specified with application of models with cubic shape voxel, with 36 μm side, corresponding to the resolution of micro-tompograph scans. In this example each voxel was filled with a single cubic finite element. The reference sample model max BV/TV had 15.7 mln DOF, and the reference model sample min BV/TV had 8.3 mln DOF.

3 Results of the Analyses

The results of numerical analyses describing the change of volumes along resolution change were presented with relative values for 3D method in Fig. 4 and FL method in Fig. 5. Along with voxel size increase there is a decrease in mineral fraction volume in the models reconstructed with both methods. Relative volume drops for 3D method exceed 20% for samples of the lowest values of BV/TV and it is only for voxels greater than 216 μm. For FL method the drops exceed 20%

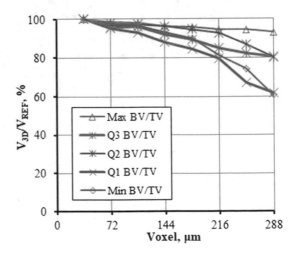

Fig. 4. Relative volume changes for method 3D

for and for all examined samples and voxels greater than 144μm. The volume
of the reconstructed structure is influenced not only by mineral fraction content
but also by trabeculae thickness. Trabeculae thicknesses for Q1 BV/TV sample
are lower than that of min BV/TV sample and Q3 BV/TV are lower than Q2
BV/TV which has an impact on relative position of appropriate graphs in Figs. 4
and 5.

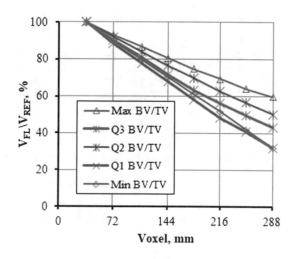

Fig. 5. Relative volume changes for method FL

The results of numerical analyses describing the change of apparent sample
modules with varied resolution are presented with relative values for 3D method
in Fig. 6 and for FL method in Fig. 7.

Along with the voxel increase there is a decrease of Young apparent modules for the samples reconstructed with both methods. Relative drops of Young
apparent modules are greater than relative drops of volume specified for both
methods. Young apparent modules for the sample is influenced not only by
mineral fraction content but also by trabeculae thickness. Contribution of bone
mineral fraction is greater for Q3 BV/TV sample than Q2 BV/TV, however trabeculae thicknesses for Q3 BV/TV are lower than Q2 BV/TV which is reflected
by higher dynamics of module decrease for Q3 BV/TV than for Q2 BV/TV.

Figure (8a) presents comparison of relative volume changes and Fig. (8b)
presents comparison of Young apparent modules obtained during numerical analyses for both methods in selected samples. Upper bar edge presents third quartile Q3 of results and lower edge presents first quartile Q1 of results. Middle line
presents the median of results. Graphs in Fig. 8 confirm the tendencies observed
for volumes in Figs. 4 and 5 and for Young apparent modules in Figs. 6 and 7.
Despite the impact of resolution, relative volumes and Young apparent modules
are underestimated.

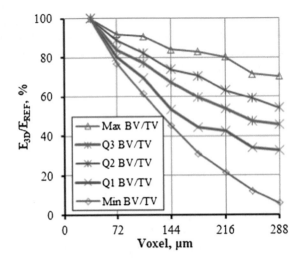

Fig. 6. Relative changes of Young apparent modules for method 3D

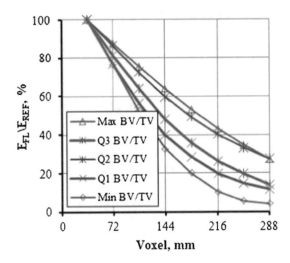

Fig. 7. Relative changes of Young apparent modules for method FL

For 3D method the median of relative volume changes is 93% and inter-quartile range is IQR = 12.6%. For FL method the median of relative volume changes is 67.8% and inter-quartile range is IQR = 26.3%. The median of volume changes for 3D method is outside the inter-quartile range for FL method. Considering the structure of extreme values of volume changes Fig. (8a) with 3D method, one can see the method has a tendency to significantly decrease this volume, whilst with FL method the extreme values are positioned symmetrically in relation to median. It means that for the selected resolution in 3D method

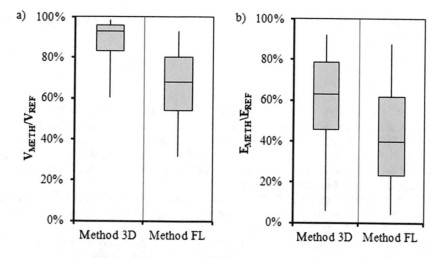

Fig. 8. Relative changes of Young apparent modules for method 3D

we are more likely to underestimate the volume and in FL method such chances are equal.

For 3D method the median of relative changes of Young apparent modules is 63.1% and inter-quartile range is IQR = 33.1%. For FL method the median of relative changes of Young apparent modules is 39.9% and inter-quartile range is IQR = 38.7%. The structure of extreme values of Young apparent modules Fig. (8b) indicates that for the selected resolution in 3D method underestimation is more likely, whilst in FL method the module overestimation is more likely.

4   Conclusions

Despite significant impact of FL method on volume decrease, its impact on Young apparent model value decrease is definitely lower. It means that FL method does not significantly change load transferring bone structure, despite its considerable influence on volume.

Numerical analyses performed with FL method provide similar information about mechanical properties of the reconstructed structure based on calculations performed for the models of lower size. Thus, potential diagnostic calculations can be performed in shorter time with application of less computer resources.

During calculations conducted with the proposed methods, each time both the values of volume as well as of Young apparent modules were lowered, so calculation results are of conservative nature. Application of either method as diagnostic model can lead to diagnosing a disease entity in healthy person. However, it does not lead to qualification of an ill person as a healthy one.

The methods of reconstruction of three-dimensional trabecular bone structures proposed in the work allow to reduce the number of scans and thus reduce the amount of radiation that the patient undergoes during CT scanning. This is done without losing important information about the mechanical features of the structure.

References

1. Topoliński, T., Cichański, A., Mazurkiewicz, A., Nowicki, K.: Microarchitecture parameters describe bone structure and its strength better than BMD. Sci. World J. **502781**, 1–7 (2012). https://doi.org/10.1100/2012/502781. Article ID 502781
2. Depalle, B., Chapurlat, R., Walter-Le-Berre, H., Bou-Said, B., Follet, H.: Finite element dependence of stress evaluation for human trabecular bone. J. Mech. Behav. Biomed. Mater. **18**, 200–212 (2013)
3. Ulrich, D., van Rietbergen, B., Weinans, H., Ruegsegger, P.: Finite element analysis of trabecular bone structure: a comparison of image-based meshing techniques. J. Biomech. **31**(12), 1187–1192 (1998)
4. Boyd, S.K., Muller, R.: Smooth surface meshing for automated finite element model generation from 3D image data. J. Biomech. **39**(7), 1287–1295 (2006)
5. Topoliński, T., Cichański, A., Mazurkiewicz, A., Nowicki, K.: The relationship between trabecular bone structure modeling methods and the elastic modulus as calculated by FEM. Sci. World J. **2012**, 1–9 (2012). https://doi.org/10.1100/2012/827196. Article ID 827196
6. Harrison, N.M., McDonnell, P.F., OMahoney, D.C., Kennedy, O.D., OBrien, F.J., McHugh, P.E.: Heterogeneous linear elastic trabecular bone modelling using micro-CT attenuation data and experimentally measured heterogeneous tissue properties. J. Biomech. **41**(11), 2589–2596 (2008)
7. Bardyn, T., Reyes, M., Larrea, X., Buchler, P.: Influence of Smoothing on Voxel-Based Mesh Accuracy in Micro-Finite Element. Springer, New York (2010)
8. Cichański, A., Nowicki, K.: Morphological dilation as the method of mineral fraction loss compensation in reconstruction of trabecular bone structure. In: Proceedings of 22nd International Conference on Engineering Mechanics 2016, pp. 114–117 (2016)
9. Cichański, A., Nowicki, K., Mazurkiewicz, A., Topoliński, T.: Investigation of statistical relationships between quantities describing bone architecture, its fractal dimensions and mechanical properties. Acta Bioeng. Biomech. **12**(4), 69–77 (2010)
10. Boutroy, S., Van Rietbergen, B., Sornay-Rendu, E., Munoz, F., Bouxsein, M.L., Delmas, P.D.: Finite element analysis based on in vivo HR-pQCT images of the distal radius is associated with wrist fracture in postmenopausal women. J. Bone Miner. Res. **23**(3), 392–399 (2008)
11. Bevill, G., Keaveny, T.M.: Trabecular bone strength predictions using finite element analysis of micro-scale images at limited spatial resolution. Bone **44**(4), 579–584 (2009)

Sport Biomechanics and Technology

Analysis of Skeletal Muscle System Loads for the Most Optimal Positions During Lifting in Different Load Distances

Bieniek Andrzej[1]([✉]), Szczygioł Anna[2], Michnik Robert[1], Chrzan Miłosz[1], Wodarski Piotr[1], and Jurkojć Jacek[1]

[1] Department of Biomechatronics, Faculty of Biomedical Engineering, Silesian University of Technology, ul. Roosvelta 40, 41-800 Zabrze, Poland
andrzej.bieniek@polsl.pl
[2] The Jerzy Kukuczka Academy of Physical Education in Katowice, ul. Mikołowska 72a, 40-065 Katowice, Poland

Abstract. The aim of this study was to determine the effect of the distance between load and the ankle joint on musculoskeletal system loading. The Any-Body software with the verified model was used for calculations of loads of muscoskeletal system during the initial phase of lifting. A total of 3,485 static musculoskeletal models in different positions were analyzed, out of which 13 with optimal lumbar spine loads were selected. Recived data from model calculation were knee joint reactions, L5S1 intervertebral disc reactions and sum of squares of muscle forces. Results confirm that the musculoskeletal system loading increase with growth of the load distance. However, it is worth to notice that optimal models basing on reactions in lumbar spine are not optimal in terms of knee joint loads and energy expenditure. In addition, there was also no change in the reactions observed in the literature for a load distance of about 0.4 m. It indicates that this change may be the result of the habits of the subjects but not the actual increase in efficiency. These study is an introduction to a broader analysis of the presented issue.

Keywords: Lifting · Inverse dynamics · Digital human modelling
Muscle force · Optimization

1 Introduction

Lifting objects is one of the most often used pattern during physical work. Moreover, it highly burden the skeletal muscular system so that it appears as an important issue in context of ergonomy. Additionally, loads arising in the lumbar spine contribute to the fact that the lifiting objects is one of the risk factor of the low back pain [11]. The prevalence and the generated loads in the skeletal muscle system cause that the problem of lifting objects is widely analyzed by researchers. The most frequently analyzed issue related to lifting objects is the analysis of the lower spine loads. Authors undertook direct measurements using

K. Arkusz et al. (Eds.): BIOMECHANICS 2018, AISC 831, pp. 221–230, 2019.
https://doi.org/10.1007/978-3-319-97286-2_20

measuring implants [12]. Attempts were also made to determine these loads by indirect methods such as modeling [2], using EMG as a support for the modeling method [16, 27], using artificial neural networks [1], or regression tools to try to estimate these loads by generated equations [4]. There are studies which aim is to compare indivudal methods [31]. Another large area of interest for researchers is to study various techniques for lifting objects. The most common topic of these analyzes is the comparison of two basic techniques, ie stoop and squat. In the literature methods of modeling the kinematics of these techniques [10] likewise the loads occurring in the lumbar spine were determined [12]. We found also one review paper [11]. Other area of reseachers' interest is calculation of the optimal way of lifting [7], determining the differences between lifting two objects on the sides and front of the body [14] as well as determining differences in kinematics raised between experienced and inexperienced workers [29] and between men and women [30]. Furthermore, the effect of distraction on the musculoskeletal system loads [23], changes in the technique of lifting for a long period of lifting [15] and the influence of limitation of mobility of the lower limbs on kinematics and musculoskeletal system loading were also examined [21]. Nevertheless, there are just few papers describing the influence of the distance of a lifiting object on its characteristics. The authors so far defined the influence of this distance on the kinematics of lifting [6], lumbar spine load [32] and the maximum mass allowed for lifting [8, 9] but there is lack of papares analyzing the entire human body while lifiting objects. Until now, there were no tools that would allow for effective analyses of the skeletal muscle system loads in the whole body lifiting task. However, along with the development of advanced biomechanical models and algorithms calculating muscle forces likewise development of computer hardware able to solve complex computational tasks, the modeling methods can be more widely used for the skeletal muscle system loading determination. The usefulness of these tools is confirmed by scientific papers [17, 28]. Another emerging tools that should be considered in the future are virtual reality technologies, which utility for ergonomic is proven in previous researches [19]. This issue is widely analyzed by researchers both in terms of the impact of virtual reality on human [25, 26] and the use of this technology in various fields [18, 22]. Using virtual reality technologie sit is possible to prototype workplaces in a way that minimizes the burden of the skeletal muscle system [19]. The aim of this study was to examine the impact of the distance of the lifted object on the musculoskeletal system loading using the biomechanical model of the whole body. According to litearature, the value of the resultant reaction in the spine at the level of L5S1 was analyzed. Additionally, after consulatation wih physiotherapists the value of the reaction in the knee joint was obtained as well as the sum of squares of muscle forces, which corresponds to the energy expenditure.

2   Material and Methods

In order to determine the loads of the skeletal muscle system during the lifting, a series of model calculations were performed. The AnyBody software was used to compute muscle foreces with the model method and the model of the whole body was delivered by the software. Given model has been positively verified many times, which can be found in the literature [5,31], as well as on the basis of own research. This model has a module that allows estimating the ground reaction forces, which enable calculations without biomechanical studies. Such proccessing allowed to generate models for regularly distributed values of angles in the joints. The used algorithm is described in the following parts of this study. Further advantages of adopted approach were the ability to examine a larger number of models, characterized by a much greater variability than in the case of testing on living humans. It was possible to investigate unmanageable positions for the majority of subjects, as well as generating significant overloads for the human body. The individual habits of the subjects were also not taken into account, which in traditional conditions is only possible for a very large and diverse research group. The results were not affected by the variability of the anthropometric data of the subjects, as all models were calculated for an identical model. Due to big number of models, analyzes were made only for the moment when the object was kept just above the ground without taking into account the accelerations accompanying the commencement of lifting, as well as its further part. The load was 5 kg, placed in each hand.

2.1   Generating Kinematic Data for Models

The input kinematic data for model were values of flexion angles in the hip, knee, elbow and lumbar spine. The flexion values for the lumbar spine was divided into joints between the individual vertebrae. The design of the model prevent direct control of the flexion angle in the ankle joint. The change in ankle flexion is achieved by changing the Center of Pressure (COP) position. In order to simplify data analysis and reduce the number of models, it was assumed that the movement takes place only in the sagittal plane. This involves maintaining neutral values for the remaining angles in the joints. As part of the preparation of kinematic data for the model, the range of variation of angles in the joints was determined and a grid of evenly spaced points with coordinates determined by the values of these angles was generated. The characteristics of the grid are shown in Table 1.

Table 1. The input range of variation value for model

	min	max	step
Hip [deg]	0	130	10
Knee [deg]	0	150	10
Spine [deg]	−30	100	10
Arm [deg]	0	100	10
COP [m]	0	0,3	0,05

In this way, 241,472 models were obtained. For each of these points, the final positions of the hand were determined. A number of models was limited to the ones for which the hand position was in the area of interest. Such proccess allowed to limit the time needed to carry out the calculations. A total of 3485 models were obtained. Sample model is shown in the figure (see Fig. 1). Muscle forces and reaction values in the joints were determined. After analyzing the levels of muscle activation, models for which the forces generated by the muscles exceeded the maximum force of the muscle were eliminated. Finally, 2679 models were obtained and analyzed. 13 points with the lowest reaction values in the L5S1 disk for the subsequent values of the distance of the load from the ankle joint were obtained (see Fig. 2). Then these points are marked on the graphs showing the relationship of the resultant reaction in the knee joint (see Fig. 3) and the sum of squares of muscle forces (see Fig. 4) as a function of the distance of the load from the ankle joint.

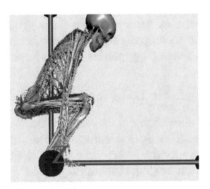

Fig. 1. Sample model.

2.2   Data Analysis

We obtained reactions and moments of forces in joints for the whole body (126 positions) and muscle forces values for individual muscle acts (816 positions). According to literature review, consultation with physiotherapists and sport coaches, it was decided to analyze the values of the resultant reactions in the L5S1 intervertebral disc, resultant reaction in the knee joint (as the most loaded joint of the lower limb during lifting) and the sum of squares of muscle forces, which is the goal function and informing about the energy expenditure of the muscles. Subsequently, optimal lifting positions for different load distances from the ankle joint were selected. For this purpose, the relationship between the resultant reaction in the intervertebral disc L5S1 as a function of the distance of the load from the ankle joint was drawn. From the obtained points, the ones lying on the curve limiting the set from the bottom was selected. The relationship of the resultant reaction in the intervertebral disc L5S1 as a function of the distance of the load from the ankle joint and the linear regression line was determined for all points excluding two points for the lowest values of the distance of the load from the ankle joint (see Fig. 5). Points these deviated from the linear characterization of the remaining points as noticed in literature review [32]. In addition, the quadratic regression curve for the entire data set was adjusted.

3   Results

The relationship of loads in the spine at the level of L5S1 was plotted as a points' cloud (see Fig. 2), positions selected for further analysis are marked black. It is observed that the remaining points have an irregular distribution and do not show a clear relationship on the load distance. The relationship of the resultant value of reaction in the knee joint as a function of the distance of the load from the ankle joint was presented as a cloud of points with marked points (see Fig. 3). It can be seen that the analyzed points remain high reaction values in the knee joint and the remaining points do not show a clear relationship with the distance the load. The relationship of the value of the sum of squares of muscle forces as a function of the distance of the load from the ankle joint is presented as a cloud of points with marked points (see Fig. 4). It can be observed that the analyzed points assume high values of the sum of squares of muscle forces and the remaining points show a certain regularity. In order to determine the characteristics of the reaction values in the intervertebral disc L5S1, the relationship of the reaction values in the spine and the regression curves were determined depending on the load distance (see Fig. 5). Equations of regression curves and corresponding R2 values are shown in the graph. The regression line is marked with a solid line as means of a quadratic function, with dashed line

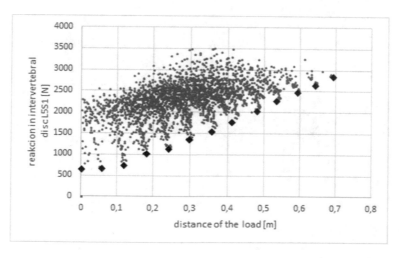

Fig. 2. Relationship of loads in intervertebral disc L5S1 as a function of the distance of the load from the ankle joint, black squares point at 13 optimal models in term of calculation of lumbar spine reactions.

linear regressions is marked. Regression and regression descriptions for the whole set of points are marked in gray, while in black - for a set limited by not taking into account points for low load distances.

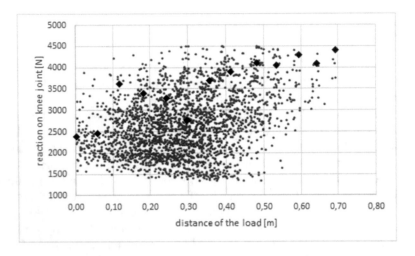

Fig. 3. Relationship of reaction on knee joint as function of the distance of the load from the ankle joint, black squares point at 13 optimal models in term of calculation of lumbar spine reactions.

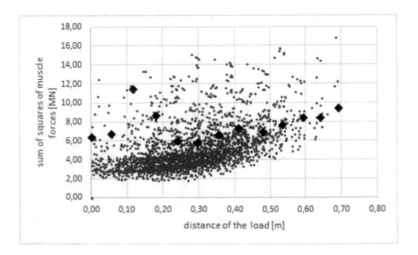

Fig. 4. The relationship of the value of the sum of squares of muscle forces as a function of the distance of the load from the ankle joint, black squares point at 13 optimal models in term of calculation of lumbar spine reactions.

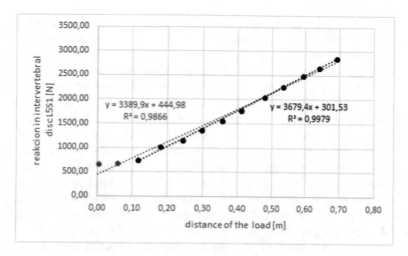

Fig. 5. Relationship of loads in intervertebral disc L5S1 as a function of the distance of the load from the ankle joint, with regression curves.

4   Discussion

The obtained results are consistent with the values obtained by researchers in previous studies [12,29]. There was an increase in the values of the resultant reactions in the spine at the level of L5S1, what is also consistent with literature [6,32]. However, there was no change [32] in the slope of the relationship of the resultant reaction in the L5S1 intervertebral disc for the distance from the ankle joint at 0.4 m. A strongly linear relationship of the reaction values in the L5S1 disc was observed as a function of the distance of the load from the ankle joint (see Fig. 5), even without elimination the points corresponding to the lowest distances. The R2 values were obtained at the level of 0.9866, which corresponds to a very good adjustment of points to the curve, ie a strongly linear character of the analyzed relationship. If the points corresponding to lowest values of the load distance are eliminated, the R2 value is at the level 0.9979 which proves that for the load distance less than 0.1 m of the resultant value of the reaction in the L5S1 intervertebral disc significantly deviate from the linear characteristics of the reaction value for distance loads higher than 0.1 m. It indicates that for optimal lifting positions, due to the resultant reaction in the L5S1 disc, there is no change in the lifting technique which was observed by [32]. Changing the lifting technique for distances over 0.4 m may result from 3 factors: the habits of the subject, increase in reaction in other joints or from a significant increase in energy expenditure. However, there were no significant increases in both knee joint reaction values (see Fig. 3) and sum of squares of muscle forces (see Fig. 4) for a load distance nearly 0.4 m. Nevertheless, it can be noted that optimal positions for a given load distance in terms of resultant reaction in the L5S1 intervertebral disc cause a significant increase in the knee joint reaction (see Fig. 3) and muscle energy expenditure (see Fig. 4). In addition, the regression was examined by means of a quadratic function, obtaining a R2 value at the level of 0.9962, which is a surprisingly high result. It can be assumed that this is only a local phenomenon and a consequence of the deviation from results for low values of the load distance. However this problem requires further calculations for models with a greater load distance from the ankle which were not analyzed in the current study. Researchers studying the biomechanics of lifting objects are mainly focused on the loads generated in the spine, especially in its lumbar unit [11,30]. They also often do not take into account the participation of muscle forces [13,20,29,33], which results in a significant underestimation of the values of musculoskeletal system loads and finally, negligence of important areas in which injuries may occur. From an ergonomic point of view, the aim of researches should not be only optimization of the time, way of lifting [7], kinematics [29] [10,24], or lumbar spine loads [3,33], but also a more comprehensive approach to load limitations generated in other sensitive elements of the skeletal muscle system as well as the reduction of energy expenditure related to the performance of tasks. This problem requires an analytical approach. Firstly, determination the skeletal muscle system loads as a function of kinematic data representing input data to the model and then calculating the optimal positions defined as continuous rather than discrete functions. Mentioned approach should give a significant improvement in the obtained results. Additionally, as reported by [15], static models are underestimated from 19% to 52% in relation to dynamic models, taking into account system accelerations and generated thereby additional external loads which are increasing both, the reaction values in the joints and the values of muscle forces. Analysis of accelerations needed to start lifting an object from the ground is one of the planned further direction of research.

References

1. Arjmand, N., Ekrami, O., Shirazi-Adl, A., Plamondon, A., Parnianpour, M.: Relative performances of artificial neural network and regression mapping tools in evaluation of spinal loads and muscle forces during static lifting. J. Biomech. **46**, 1454–1462 (2013). https://doi.org/10.1016/j.jbiomech.2013.02.026
2. Arjmand, N., Gagnon, D., Plamondon, A., Shirazi-Adl, A., Larivière, C.: A comparative study of two trunk biomechanical models under symmetric and asymmetric loadings. J. Biomech. **43**, 485–491 (2010). https://doi.org/10.1016/j.jbiomech. 2009.09.032
3. Arjmand, N., Plamondon, A., Shirazi-Adl, A., Larivière, C., Parnianpour, M.: Predictive equations to estimate spinal loads in symmetric lifting tasks. J. Biomech. **44**, 84–91 (2011). https://doi.org/10.1016/j.jbiomech.2010.08.028
4. Arjmand, N., Plamondon, A., Shirazi-Adl, A., Parnianpour, M., Larivière, C.: Predictive equations for lumbar spine loads in load-dependent asymmetric one- and two-handed lifting activities. Clin. Biomech. **27**, 537–544 (2012). https://doi.org/ 10.1016/j.clinbiomech.2011.12.015
5. Bassani, T., Stucovitz, E., Qian, Z., Briguglio, M., Galbusera, F.: Validation of the AnyBody full body musculoskeletal model in computing lumbar spine loads at L4L5 level. J Biomech. (2017). https://doi.org/10.1016/j.jbiomech.2017.04.025

6. Burgess-Limerick, R., Abernethy, B.: Effect of lead distance on self-selected manual lifting technique. Int. J. Ind. Ergon. **22**, 367–372 (1998). https://doi.org/10.1016/S0169-8141(97)00090-5

7. Chang, C.C., Brown, D.R., Bloswick, D.S., Hsiang, S.M.: Biomechanical simulation of manual lifting using spacetime optimization. J. Biomech. **34**, 527–532 (2001). https://doi.org/10.1016/S0021-9290(00)00222-0

8. Ciriello, V.M.: The effects of box size, frequency and extended horizontal reach on maximum acceptable weights of lifting. Int. J. Ind. Ergon. **32**, 115–120 (2003). https://doi.org/10.1016/S0169-8141(03)00045-3

9. Ciriello, V.M.: The effects of container size, frequency and extended horizontal reach on maximum acceptable weights of lifting for female industrial workers. Appl. Ergon. **38**, 1–5 (2007). https://doi.org/10.1016/j.apergo.2006.02.001

10. Colobert, B., Multon, F., Cretual, A., Delamarche, P.: Biomechanical simulation of human lifting. In: ESM 2003: 17th European Simulation Multiconference: Foundations for Successful Modelling and Simulation, pp. 318–322 (2003)

11. Van Dieën, J.H., Hoozemans, M.J.M., Toussaint, H.M.: Stoop or squat: a review of biomechanical studies on lifting technique. Clin. Biomech. **14**, 685–696 (1999)

12. Dreischarf, M., Rohlmann, A., Graichen, F., Bergmann, G., Schmidt, H.: In vivo loads on a vertebral body replacement during different lifting techniques. J. Biomech. **49**, 890–895 (2016). https://doi.org/10.1016/j.jbiomech.2015.09.034

13. Faber, G.S., Chang, C.C., Kingma, I., Dennerlein, J.T.: Estimating dynamic external hand forces during manual materials handling based on ground reaction forces and body segment accelerations. J. Biomech. **46**, 2736–2740 (2013). https://doi.org/10.1016/j.jbiomech.2013.07.030

14. Faber, G.S., Kingma, I., Bakker, A.J.M., van Dieën, J.H.: Low-back loading in lifting two loads beside the body compared to lifting one load in front of the body. J. Biomech. **42**, 35–41 (2009). https://doi.org/10.1016/j.jbiomech.2008.10.013

15. Fogleman, M., Smith, J.L.: The use of biomechanical measures in the investigation of changes in lifting strategies over extended periods. Int. J. Ind. Ergon. **16**, 57–71 (1995). https://doi.org/10.1016/0169-8141(94)00087-J

16. Gagnon, D., Larivière, C., Loisel, P.: Comparative ability of EMG, optimization, and hybrid modelling approaches to predict trunk muscle forces and lumbar spine loading during dynamic sagittal plane lifting. Clin. Biomech. **16**, 359–372 (2001). https://doi.org/10.1016/S0268-0033(01)00016-X

17. Guzik-Kopyto, A., Michnik, R., Wodarski, P., Chuchnowska, I.: Determination of loads in the joints of the upper limb during activities of daily living. In: Advances in Intelligent Systems and Computing, pp. 99–108 (2016). https://doi.org/10.1007/978-3-319-39904-1_9

18. Gzik, M., Wodarski, P., Jurkojć, J., Michnik, R., Bieniek, A.: Interactive system of enginering support of upper limb diagnosis. In: Advances in Intelligent Systems and Computing, pp. 115–123 (2017)

19. Hu, B., Ma, L., Zhang, W., Salvendy, G., Chablat, D., Bennis, F.: Predicting real-world ergonomic measurements by simulation in a virtual environment. Int. J. Ind. Ergon. **41**, 64–71 (2011). https://doi.org/10.1016/j.ergon.2010.10.001

20. Hwang, S., Kim, Y., Kim, Y.: Lower extremity joint kinetics and lumbar curvature during squat and stoop lifting. BMC Musculoskelet. Disord. (2009). https://doi.org/10.1186/1471-2474-10-15

21. Jin, S., Mirka, G.A.: The effect of a lower extremity kinematic constraint on lifting biomechanics. Appl. Ergon. **42**, 867–872 (2011). https://doi.org/10.1016/j.apergo.2011.02.003

22. Jurkojć, J., Wodarski, P., Michnik, R., Nowakowska, K.: The upper limb motion deviation index: a new comprehensive index of upper limb motion pathology. Acta Bioeng. Biomech. **19**, 175–185 (2016). https://doi.org/10.5277/ABB-00698-2016-02

23. Katsuhira, J., Matsudaira, K., Iwakiri, K., Kimura, Y., Ohashi, T., Ono, R., Sugita, S., Fukuda, K., Abe, S., Maruyama, H.: Effect of mental processing on low back load while lifting an object. Spine **38**, E832–E839 (2013). https://doi.org/10.1097/BRS.0b013e31829360e5. (Phila Pa 1976)

24. Lee, J., Nussbaum, M.A.: Experienced workers may sacrifice peak torso kinematics/kinetics for enhanced balance/stability during repetitive lifting. J. Biomech. **46**, 1211–1215 (2013). https://doi.org/10.1016/j.jbiomech.2013.01.011

25. Michnik, R., Jurkojć, J., Wodarski, P., Gzik, M., Bieniek, A.: The influence of the scenery and the amplitude of visual disturbances in the virtual reality on the maintaining the balance. Arch. Budo **10**, 133–140 (2014)

26. Michnik, R., Jurkojć, J., Wodarski, P., Gzik, M., Jochymczyk-Woźniak, K., Bieniek, A.: The influence of frequency of visual disorders on stabilographic parameters. Acta Bioeng. Biomech. **18**, 25–33 (2016). https://doi.org/10.5277/ABB-00201-2014-04

27. Mohammadi, Y., Arjmand, N., Shirazi-Adl, A.: Comparison of trunk muscle forces, spinal loads and stability estimated by one stability- and three EMG-assisted optimization approaches. Med. Eng. Phys. **37**, 792–800 (2015). https://doi.org/10.1016/j.medengphy.2015.05.018

28. Nowakowska, K., Gzik, M., Michnik, R., Myśliwiec, A., Jurkojć, J., Suchoń, S., Burkacki, M.: Innovations in Biomedical Engineering. Springer, Cham (2017)

29. Plamondon, A., Delisle, A., Bellefeuille, S., Denis, D., Gagnon, D., Larivière, C., IRSST MMH Research Group: Lifting strategies of expert and novice workers during a repetitive palletizing task. Appl. Ergon. **45**, 471–481 (2014). https://doi.org/10.1016/j.apergo.2013.06.008

30. Plamondon, A., Larivière, C., Denis, D., St-Vincent, M., Delisle, A.: Sex differences in lifting strategies during a repetitive palletizing task. Appl. Ergon. **45**, 1558–1569 (2014). https://doi.org/10.1016/j.apergo.2014.05.005

31. Rajaee, M.A., Arjmand, N., Shirazi-Adl, A., Plamondon, A., Schmidt, H.: Comparative evaluation of six quantitative lifting tools to estimate spine loads during static activities. Appl. Ergon. **48**, 22–32 (2015). https://doi.org/10.1016/j.apergo.2014.11.002

32. Schipplein, O.D., Reinsel, T.E., Andersson, G.B., Lavender, S.A.: The influence of initial horizontal weight placement on the loads at the lumbar spine while lifting. Spine **20**, 1895–1898 (1995). https://doi.org/10.1097/00007632-199509000-00010. (Phila Pa 1976)

33. Visser, S., Faber, G.S., Hoozemans, M.J.M., van der Molen, H.F., Kuijer, P.P.F.M., Frings-Dresen, M.H.W., van Dieën, J.H.: Lumbar compression forces while lifting and carrying with two and four workers. Appl. Ergon. **50**, 56–61 (2015). https://doi.org/10.1016/j.apergo.2015.02.007

Strength, Flexibility and Temperature Changes During Step Aerobics Training

Piotr Borkowski[1]([✉]), Jolanta Grażyna Zuzda[2],
and Robert Latosiewicz[3]

[1] Faculty of Mechanical Engineering, Department of Biocybernetics
and Biomedical Engineering, Bialystok University of Technology,
Wiejska 45, 15-351 Bialystok, Poland
p.borkowski@pb.edu.pl
[2] Tourism and Recreation Department, Bialystok University of Technology,
Tarasiuka 2, 16-001 Bialystok, Poland
j.zuzda@pb.edu.pl
[3] Faculty of Health Sciences, Department of Rehabilitation and Physiotherapy,
Medical University of Lublin, Magnoliowa 2, 20-143 Lublin, Poland
r.latosiewicz@umlub.pl

Abstract. Purpose: The purpose of this study was the analysis of temperature changes in the muscles quadriceps and biceps femoris areas of healthy subjects over a 15-week Step Aerobics Training (SAT) program. The aim of this paper is twofold. Firstly we verify if SAT training has an impact on the temperature changes in the muscles quadriceps and biceps femoris and thermoregulation of healthy subjects. Secondly, we verify if there is relationship between the index of strength and flexibility with thermal results.

Methods: The study was conducted with 11 women subjects aged between 20- and 22-years-old. Training sessions of SAT took 15 weeks. Subjects performed two training session per week for 60 min per session. Assessments included the evaluation of body mass index, waist circumference, blood pressure, vital capacity of lungs, maximal oxygen uptake and flexibility and strength. Subjects' fitness was evaluated by testing the following tasks: standing long jumps and toe-touch test. For thermograms acquisition a thermographic camera CEDIP Titanium 560M IR (USA) was used.

Results: After 15 weeks SAT the temperature changes were found to be larger on the front surfaces than on back surfaces of lower extremities. Positive correlation between strength and skin temperature and negative correlation between flexibility and temperature was found.

Conclusions: 15-weeks-long SAT can promote improvements in the thermoregulation of apparently healthy women. This method may be used as a tool for establishing the efficiency of SAT in the training process.

Keywords: Thermal images · Thermoregulation step aerobics training
Strength · Flexibility

K. Arkusz et al. (Eds.): BIOMECHANICS 2018, AISC 831, pp. 231–241, 2019.
https://doi.org/10.1007/978-3-319-97286-2_21

1 Introduction

Physical activity (PA) has major beneficial effects on the general health of the population worldwide. SAT has been recommended one of the most popular collective forms of fitness. SAT may have a positive impact on the human body. It is recommended for the improvement of the cardio respiratory and muscular fitness [7, 21, 25].

During PA as a result of muscle contraction chemical energy is transformed into heat. This process increases local metabolic activity and perfusion. thus leading to a rise in the temperature of the muscle. Measuring energetic - metabolic activity of muscles presents a significant technical problem [18].

Infrared thermography IRT has been defined as the scientific analysis of data from non-contact, safe and low cost thermal imaging devices [5, 13, 18]. Thermography is one non-invasive method used to measure the thermal radiation emitted by the body or its parts and can be used to identify temperature changes caused by PA. Subsequently infrared thermography has become a powerful investigative tool with many applications: mechanical, electrical, military, building, medical and few studies using infrared thermography have been devoted to sports performance and sports pathology diagnostic based on its non-invasiveness [2, 4, 12, 14, 15, 18, 24, 26].

Although there are few investigations that have confirmed beneficial health outcomes associated with SAT [16, 17, 21, 22]. These studies have not evaluated its effects on the temperature of the muscles quadriceps and biceps femoris. Therefore, we designed this study with a twofold aim. Firstly, to verify if SAT has an impact on temperature changes in the in the muscles quadriceps and biceps femoris. Secondly, to verify if there is any relationship between the index of strength and flexibility with thermal results.

2 Materials and Methods

The research was performed on 11 healthy female volunteers. The participants of mean age 21.0. ± 0.45 years and height 1.65. ± 00.07 cm were identified and randomly selected from Bialystok University of Technology Poland. All subjects were examined by a medical doctor pre-training 15 weeks post-training. The PAR-Q was used to define inclusion/exclusion criteria. Before training all subjects were informed about the training protocol and gave informed consent regarding study procedures. Inclusion criteria for subjects recruited for the study included their general health status (to determine that they would be capable of completing the training programme) and non-participation in any SAT course in the 15-weeks preceding the study. The subjects participated in the study with their consent in accordance with the Declaration of Helsinki and the approval of the Ethical Committee.

2.1 Methods

Each subject's body weight was measured using a scale with a resolution of 100 g. The subject's height was measured by stadiometer WPT 100/200 OW. Subjects' musculoskeletal fitness and flexibility were evaluated by testing the following tasks: standing long jumps and a toe-touch test [20].

For thermograms acquisition a thermographic camera CEDIP Titanium 560M IR (USA) was used. The integrated resolution of the camera was 640×512 pixels with an accuracy to 1 °C of absolute temperature. The camera was placed perpendicular to the scanned surface. Thermographic images were taken in a standing position in an air–conditioned room with a constant temperature of 23 °C. For each individual two series of 4 sequenced body thermograms were taken: 2 from the front of the body and 2 from the back. The quadriceps femoris right and left (QfR, QfL) and biceps femoris right and left (BfR, BfL) were subjected to automatic processing and calculation.

The thermograms were taken at the following four times: at the first session after 10 min of warm up, after the first session of training, at the last session after 10 min of warm up and after the last session of training. Thermograms were analyzed using original 5.80.005 software (Altair Engineering. Inc. USA).

2.2 Step Aerobics Training

SAT was conducted according to valid collective fitness forms of training methodology [25]. Training sessions lasted for 15 weeks and were conducted two times a week on non-consecutive days and were led by the same certified instructor (JZ) under standardized conditions regarding artificial light, air temperature 23 °C and relative humidity of 50%. Each training session of SAT consisted of the following parts: warm-up, main training activity and cool-down. The intensity during the sessions was controlled by music tempo and oscillated between 120 BPM to 145 BPM. Licensed "workout music" was used.

2.3 Statistical Analysis

Means and standard deviations were calculated for the demographic data of all participants. For basic data characteristics the Shapiro-Wilk test was used for the analysis of normal distribution. The data from the three thermograms, taken at each of the four times outlined above, were averaged to provide a unique score for each participant. The temperature changes between pre-training and post-training were compared using the Wilcoxon test. The relationship between the index of waist circumference, strength and flexibility with thermal results was assessed by using a non-parametric correlation test. A value of $p < 0.05$ was considered significant. Computer software Statistica 12.5 (StatSoft Polska, Poland) was used for analysis.

3 Results

3.1 Subject Characteristic

The mean age of all subjects at the beginning of the study was 21.0 ± 0.45 years old and height 1.65 ± 0.07 cm. Characteristics of participants at pre-1st SAT and post-15th SAT are shown in Table 1.

Table 1. Skin areas of the selected muscles, their parts or groups

Skin areas abbreviations	Muscles, their parts or groups
QfR	Quadriceps femoris muscle right
QfL	Quadriceps femoris muscle left
BfR	Biceps femoris muscles right
BfL	Biceps femoris muscles left

After 15 weeks of SAT the mean value of participant BMI declined to 22.48 ± 1.83 kg/m^2 (p > 0.05). The mean value of waist circumference (WC) also reduced to 75.09 ± 7.71 cm (p < 0.05). The mean value of VO$_2$max was almost stable (39.20 ± 5.85 ml/kg/min vs. 39.97 ± 5.78 ml/kg/min. p > 0.05. Speed-power index (measured by the length of the standing jump) pre-training was 1.53 ± 0.15 m and was a higher post-training: 1.59 ± 0.15 m (p < 0.05). The average value of the flexibility index (touch-toe test) after 15 weeks of SAT increased to 11.77 ± 5.89 cm as compared to pre-training 10.55 ± 5.3 cm (p < 0.05). A significant difference in strength and flexibility between pre-training and post-training levels was observed for subjects (p < 0.05; Table 2).

Table 2. Female characteristics pre-training and 15 weeks post-training (n = 11)

Subject's characteristic	Pre-1SAT Mean (SD)	Pro15SAT Mean (SD)	Z	P
BMI [kg/m^2]	22.67 (1.72)	22.48 (1.83)	1.95	0.05
Waist circumference [cm]	77.36 (6.79)	75.09 (7.71)	2.24	0.03*
Systolic blood pressure [mm Hg]	121.4 (10.3)	126. 1 (11.6)	1.48	0.14
Diastolic blood pressure [mm Hg]	72.1 (9.4)	73.4 (9.5)	0.24	0.81
Vital capacity of lungs [ml]	3854.6 (738.1)	3918.2 (744.1)	0.631	0.53
VO$_2$max [ml/kg/min]	39.20 (5.85)	39.97 (5.78)	0.89	0.38
Weight [kg][m]	61.98 (5.33)	61.46 (5.33)	1.96	0.05
Jump [m]	1.53 (0.15)	1.59 (0.15)	2.8	0.005*
Flexibility [cm]	10.55 (5.3)	11.77 (5.89)	2.52	0.01*

3.2 Thermal Images Assessment

The temperature of the region of the muscles quadriceps and biceps femoris of the legs was registered: at the first session after 10 min of warm up, after the first session of training, at the last session after 10 min of warm up and after the last session of training. Examples of thermograms of selected areas: QfR, QfL are presented in Fig. 1.

Pre-1SAT a)	Post - 1SAT. b)

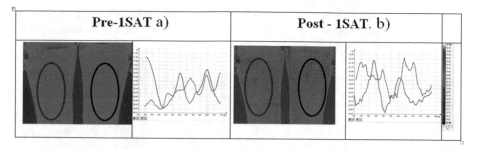

Fig. 1. Examples of pre-training and post-training thermal images (from left to right): (a) before first SAT, (b) after first SAT

3.3   Analysis of Temperature Changes

Medians of temperatures of QfL an QfR after 1st SAT were higher than before 1st SAT (Table 3). Medians of temperatures QfR increased by 0,6 °C (e.g. 29,6 °C to 30,2 °C; $p < 0.05$) and QfL increased by 1.08 °C (e.g. 29,59 °C to 30,29 °C; $p < 0.05$). Medians of temperatures BfR decreased by 0,5 °C (e.g. 30,4 °C to 29,9 °C; $p < 0.05$) and BfL decreased by 1.5 °C (e.g. 30,58 °C to 29,08 °C; $p < 0.05$).

Table 3. Descriptive statistics of skin temperature (°C) in evaluated muscles areas before and after of SAT (n = 11)

Muscle	Before/after	1 SAT			15 SAT			1 SAT-15 SAT			WR-L	
		mc	25%	75%	mc	25%	75%	mc	25%	75%	Z	P
QfR	Before	29,6	28,8	30,1	30,2	29,8	30,75	0,5	–0,04	1,6	76,3	0,00
	After	30,2	29,3	31.0	30,6	29,7	31,4	0,5	–0,6	1,5	85,60	0,00
QfL	Before	29,59	29,1	30,00	30,4	29,9	31,0	2,16	0,86	2,98	77,90	0,00
	After	30,29	29,26	31,13	30,75	30,03	31,51	0,56	–,39	1,51	81,93	0,00
BfR	Before	30,44	29,99	31,06	31,58	30,40	32,39	1,05	0,07	1,86	42,18	0,00
	After	29,90	29,05	30,65	30,39	29,69	31,12	0,54	–0,38	1,37	49,89	0,00
BfL	Before	30,58	29,97	31,19	31,68	30,52	32,50	0,90	0,11	1,72	48,13	0,00
	After	29,80	28,88	30,54	30,36	29,67	30,96	0,37	–0,33	1,20	46,72	0,00

Key: Mc -; Q1 – 1st quartile; Q3 – 3rd quartile; R-L right-left; W R-L – results of Willcoxon test of right – left differences; $p < 0.005$. Abbreviation of the areas are clarified in methods Table 1.

Medians of temperatures of QfL an QfR after 15 th SAT were higher than before 15 th SAT (Table 3). Medians of temperatures QfR increased by 0,3 °C (e.g. 30,2 °C to 30,5 °C; $p < 0.05$) and QfL increased by 0,35 °C (e.g. 30,4 °C to 30,75 °C; $p < 0.05$). Medians of temperatures BfR decreased by 1,19 °C (e.g. 31,58 °C to 30,39 °C; $p < 0.05$) and BfL decreased by 1.32 °C (e.g. 31,68 °C to 30,36 °C; $p < 0.05$).

After the 15 SAT (Table 3) a statistically-significant difference between the temperatures of the quadriceps femoris right and left muscles and biceps femoris right and left of subjects was found (Fig. 2).

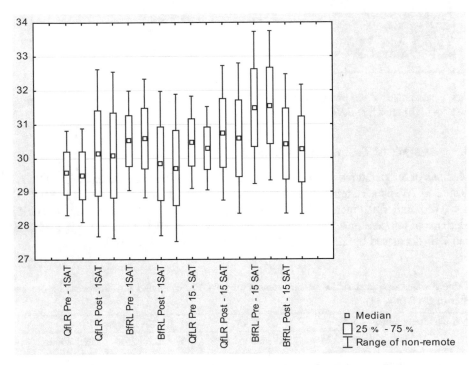

Fig. 2. Changes in temperature of individual muscles areas pre–1st SAT and post–15th SAT (key: 25% – 1st quartile; 75% – 3rd quartile. Abbreviation of the areas are clarified in Table 1)

The difference in mean prey-training 15th SAT and post-15th SAT temperature for was lower than the difference in mean pre-training 1st SAT and post-1stSAT (Table 4).

The relationship between strength and thermal results was not statistically significant but the correlation was low (r = .20398; p > 0.05) (Fig. 3). The relationship between flexibility and thermal results was not statistically significant o the correlation was low (r = −1906. p > 0.05) (Fig. 4).

Table 4. Differences in temperature before and after 60 min-long SAT (Willcoxon's test)

Area	1 SAT			15 SAT		
	Median temperature difference	W$_{A-B}$		Median temperature difference	W$_{A-B}$	
		Z	p		Z	P
QfR	0,71	46,37	0,00	0,50	32,17	0,00
QfL	0,79	47,94	0,00	0,54	31,87	0,00
BfR	−0,49	52,96	0,00	−1,15	84,22	0,00
BfL	−0,82	70,46	0,00	−1,36	100,14	0,00

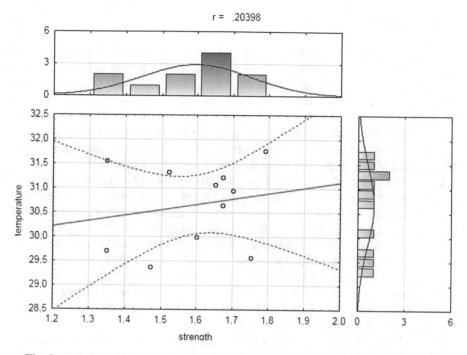

Fig. 3. Relationship between the index of strength and thermal results pre–15th SAT

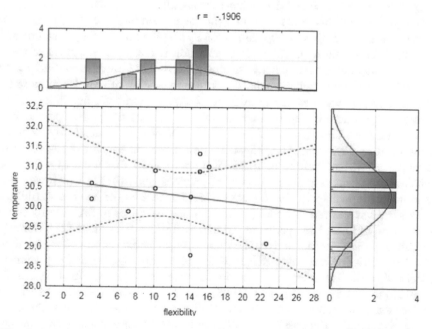

Fig. 4. Relationship between the index of flexibility and thermal results post–15th SAT

4 Discussion

Physical activity (PA) has major beneficial effects on the general health of the population worldwide. SAT may have a positive impact on the human body and has been recommended as one of the most popular collective forms of fitness. It is recommended for the improvement of the cardio–respiratory and muscular fitness [7, 11, 16, 17, 21, 26]. SAT training is associated with an array of cardiovascular, pulmonary and metabolic adaptations. Exercise has a noted effect on skin blood flow and temperature [10, 24]. The surface temperature of the body during physical activity increases and stabilizes at a higher level in proportion to the intensity of exercise [5].

Infrared thermography method is a procedure for non-contact measurement of thermal radiation on the surface of the body for monitoring various types of loading on an athlete's muscle segments and may be used as a comparative tool for establishing the efficiency of different means and methods in the training process, [1, 3, 5]. Despite the important role of temperature regulation in human behavior, it is frequently overlooked as a thermoregulatory response during both rest and exercise [23]. Infrared thermography has become a powerful investigative tool with many applications: mechanical, electrical, military, building, medical and few studies using infrared thermography have been devoted to sports performance and sports pathology diagnostic based on its noninvasiveness [2, 4, 12, 14, 15, 18, 24, 25].

In recent years the temperature changes during exercises in different areas of the body were recorded: on the quadriceps of subjects performing squat exercises, skin surface over the thigh muscle [5], surface temperature of the upper limbs [2], calves during resistance exercise and dorsal hand during office work [8].

The purpose of this study was to evaluate a usefulness of thermography and analysis of thermal images of legs during SAT as well as presenting standards of its' use in the SAT monitoring. The main explanation is that SAT is a form of training mainly involving lower parts of body (abdomen, pelvis and lower extremities) and even basic steps and choreographic combinations require activity of muscles of those body parts. Therefore, it was decided to evaluate changes in strength, flexibility and temperature changes during step aerobics training in regions of quadriceps and biceps femoris muscles. The results of our study demonstrate statistically significant improvement in flexibility. After 15-weeks period the toe–touch test results show a statistically significant raise from 10.55 (5.3) cm to 11.77 (5.89) cm (p = 0.01). Our results are consistent with the results of our previous study [41] in which we showed that people who participated in 30 weeks of SAT increased the level of flexibility.

Some studies showed that addition of resistance exercises to SAT improved muscular strength of lower body parts greater than performing SAT alone [21]. Our results demonstrate positive impact of dynamic strength of lower extremities after 15-weeks of SAT. The results of our study demonstrate statistically significant improvement in explosive force of muscles (assessed in standing long jump) which raised from 1.53 (0.15) m to 1.59 (0.15) m (p = 0.005). It is not line with the study of Drobnik–Kozakiewicz [6] who showed that 10-week SAT was no sufficient to change the strength of lower extremities. The possible explanation is that the participants were students of sport university. In our group there were females with low initial physical fitness. It is quite possible that in low-active individuals adaptive changes occur more quickly and evidently.

An important physiological advantage of training is an increase in the ability to remove heat from the body, which allows the continuation of effort. Changes in the body surface temperature may therefore indicate the loading of the locomotor system. In our study this physiologic reaction occurred in both of the examined regions, still it was prominent in the both regions of quadriceps rather than in biceps femoris regions. The temperature of quadriceps regions increased from 29.6 °C to 30.6 °C (right side) and 30.37 °C (left side). In our opinion sweating in those regions is not significantly effective in reducing heat produced by hard–working thigh muscles during SAT. On the back side of legs in the region of left biceps femoris muscles the temperature drop ranged from 30.44 °C to 30.39 °C, while on the right side drop ranged from 30,58 °C to 30,3 °C. This finding has no clear physiological explanation.

In our opinion, the 15-week SAT caused a widely–known body adaptation related to activation of the parasympathetic nervous system. It is can be assumed that exercise–induced sweating (being a physiological effect of parasympathetic function) was effective in significantly decreasing skin temperature in all but both femur regions. On the other hand, higher overall performance and stress tolerance caused by 15 weeks of SAT could also be factors that allowed better heat distribution during exercise.

We believe thermographic imaging can be used as a comparative tool for establishing the efficiency and the reaction of the skin under the influence of physical effort. Thermography measures have been increasingly use as preventive resource an control method in SAT monitoring and prevent possible injury. The detection by thermal imaging is based on standard temperature and thermal asymmetry patterns. Larger asymmetries – more than 0.60 °C may correspond to musculoskeletal injuries (normally they do not exceed values greater than 0.25 °C) [19]. Evaluating local and systemic cutaneous blood flow adaptation as a function of specific type, intensity and duration of exercise, and helping to determine the ideal intensity physical activities should be conducted in order to injury prevention strategies.

5   Conclusions

1. 15-weeks-long SAT in group of healthy young women resulted in increase of mean temperature of quadriceps muscles and decrease of mean temperature of biceps femoris muscles which should be related to improvements in the thermoregulation.
2. The positive correlation between muscle strength and the decrease of skin temperature and negative correlation between flexibility of body and skin temperature was found.
3. Thermal imaging seems to be safe and non-invasive technique of monitoring of the efficiency of physical recreational training.

References

1. Abate, M., et al.: Comparison of cutaneous termic response to a standardised warm up in trained and untrained individuals. J. Sports Med. Phys. Fit. **26**(53), 18–37 (2013)
2. Chudecka, M.: The use of thermal imaging to evaluate body temperature changes of athletes during training and a study on the impact of physiological and morphological factors on skin temperature. Hum. Mov. **13**(1), 33–39 (2012)
3. Chudecka, M.: Use of thermal imaging in the evaluation of body surface temperature in various physiological states In patients with different body compositions and varying levels of physical activity. Cent. Eur. J. Sport Sci. Med. **2**(2), 15–20 (2013)
4. Clark, R.P., Mullan, B.J., Pugh, L.G.: Skin temperature during running – a study using infrared color thermograph. J. Physiol. **267**, 53–62 (1997)
5. Coh, M.: Use of thermovision method in sport training. Phys. Educ. Sport **5**(1), 85–94 (2007)
6. Drobnik-Kozakiewicz, I., Sawczym, M., et al.: The effects of a 10-weeks step aerobics training on VO₂max, isometric strength and body composition of young women. Cent. Eur. J. Sport Sci. Med. **4**(4), 3–9 (2013)
7. Forte, R., De Vito, G., Murphy, N., et al.: Cardiovascular response during low-intensity step-aerobic dance in middle-aged subjects. Eur. J. Sport Sci. **1**(3), 1–7 (2001)
8. Gold, J.E., Cherniack, M., Buchholz, B.: Infrared thermography for examination of skin temperature in the dorsal hand of office workers. Eur. J. Appl. Physiol. **93**, 245–251 (2004)
9. Hesson, J.L.: Weight Training for Life, 9th edn., Wadsworth (2010)
10. Ludwig, N., et al.: Comparison of image analysis methods in skin temperature measurements during physical exercise. In: Quantitative InfraRed Thermography (QIRT) Conference, QIRT (2014)
11. Hallage, T., Krause, M.P., Haile, L., Miculis, C.P., Nagle, E.F., Reis, R.S.: The effects of 12 weeks of step aerobics training on functional fitness of elderly women. J. Strength Cond. Res. **24**(8), 2261–2266 (2010)
12. Hildebrandt, C., et al.: An overview of recent application of medical infrared thermography in sports medicine in Austria. Sensors **10**(5), 4700–4715 (2010)
13. Hildebrandt, C., et al.: The application of medical infrared thermography in sports medicine. In: An International Perspective on Topics in Sports Medicine and Sports Injury, pp. 257–274 (2012)
14. Hoover, K.C., Burlingame, S.E., Lautz, C.H.: Opportunities and challenges in concrete with thermal imaging. Concr. Int. **26**, 23–27 (2004)
15. Kaminski, A., Jouglar, J., Volle, C., Natalizio, S., Vuillermoz, P.L., Laugier, A.: Non-destructive characterization of defects in devices using infrared thermography. Microelectron. J. **30**, 137–1140 (1999)
16. Kravitz, L., Cisar, C.J., Christensen, C.L., Setterlund, S.S.: The physiological effects of step training with and without handweights. J. Sports Med. Phys. Fit. **33**(4), 348–358 (1993)
17. Mori, Y., Ayabe, M., Yahir, T., et al.: The effects of home-based bench step exercise on aerobic capacity. Lower extremity power and static balance in older adults. Int. J. Sport Health Sci. **4**, 570–576 (2006)
18. Novotny, J.: The influence of breaststroke swimming on the muscle activity of young men in thermographic imaging. Acta Bioeng. Biomech. **17**(2), 121–129 (2015)
19. Neves, E.B., Matos, F., Martins da Cunha, R., Reis, V.M.: Thermography to monitoring of sports training: an overview. Pan Am. J. Med. Thermol. **2**(1), 18–22 (2010)

20. Oja, P., Tuxworth, B.: Eurofit for adults. Assessment of health – relate fitness. Committee for the Development of Sport. Council of Europe. Strasbourg Cedex and UKK Institute for Health Promotion research, Tampere (2005)
21. Pereira, A., et al.: Combined strength and step aerobics training leads to significant gains in maximal strength and body composition in women. J Sports Med Phys Fitness. 53(3 Suppl. 1), 38–43 (2013)
22. Santos-Rocha, R.A., Oliveira, C.S., Veloso, A.P.: Osteogenic index of step exercise depending on choreographic movements. session duration, and stepping rate. Br. J. Sports Med. 40(10), 860–866 (2006)
23. Schlader, Z.J., Stannard, S.R., Mündel, T.: Human thermoregulatory behavior during rest and exercise - a prospective review. Physiol. Behav. 99(3), 269–275 (2010)
24. Wu, C.L., Yu, K.L., Chuang, H.Y., Huang, M.H., Chen, C.H.: The application of infrared thermography in the assessment of patients with coccygodynia before and after manual therapy combined with diathermy. J. Manip. Physiol. Ther. 14, 281–293 (2009)
25. Zuzda, J.G., Latosiewicz, R.: A Method of Conducting Recreational Classes Using the System of Reebok Step Exercises. Białystok University of Technology, Białystok (2010)
26. Zuzda, J.G., Latosiewicz, R.: Zmiana temperatury powierzchni ciała podczas rekreacyjnego uprawiania łucznictwa – badanie pilotażowe. (Changing of Body Temperature During Archery Recreation – Pilot Examination) 2, 147–158 (2010)

On Different Methods for Calculating the Flight Height in the Vertical Countermovement Jump Analysis

Jakub Krzysztof Grabski$^{(\boxtimes)}$ ⓘ, Tomasz Walczak ⓘ,
Martyna Michałowska ⓘ, Patrycja Pastusiak, and Marta Szczetyńska

Institute of Applied Mechanics, Faculty of Mechanical Engineering
and Management, Poznan University of Technology,
Jana Pawła II 24, 60-965 Poznań, Poland
jakub.grabski@put.poznan.pl

Abstract. Vertical countermovement jump is a very simple and common method for assessing the jumping ability of athletes. There are different techniques for measurements of the flight height, e.g. using motion capture systems or accelerometers. In this paper for estimating the flight height the measurements coming from the force plates are used. Furthermore different methods can be applied for calculating the flight height based on these measurements. In these paper four methods of calculating the flight height during the vertical countermovement jump based on the measurements from the force plates are compared (the flight time method, the take-off velocity method, the work-energy method and the center of jumper's body vertical position method). In addition for two of these methods (the take-off velocity method and the center of jumper's body vertical position method) the authors applied two different methods of numerical integration (the trapezoidal rule and based on the cubic spline interpolation).

Keywords: Vertical jump · Countermovement jump · Flight height
Numerical integration

1 Introduction

Vertical jump can be very helpful in assessing jumping ability of athletes. This method is very common and widely described in the literature because of its simplicity.

In biomechanics there are different techniques of measurements [1], e.g. motion capture systems [2, 3], force plates [4–6], electromyography [7, 8], dynamometers [9, 10], etc. In the literature on vertical jumps also different techniques of measurements are described. In the simplest method of measuring the vertical jump height a gauge fixed on a wall is applied [11, 12]. In recent years more and more common are methods based on measurements coming from the accelerometers [13–15]. In this study the force plates were used in order to measure the ground reaction force (GRF) during the vertical countermovement jump. Based on the GRF measurements the velocity and position of the center of the jumper's body (CJB) can be obtained by twice integration of the equations of motion.

© Springer Nature Switzerland AG 2019
K. Arkusz et al. (Eds.): BIOMECHANICS 2018, AISC 831, pp. 242–251, 2019.
https://doi.org/10.1007/978-3-319-97286-2_22

Based on the measurements coming from the force plates different methods of calculating the flight height can be applied. In this paper four different methods are compered. The flight height is defined as a difference between the peak vertical position and the vertical position at the take-off instant of the CJB. The methods compared in the paper are: the flight time method, the take-off velocity method, the work-energy method and the center of jumper's body vertical position method. Furthermore for the take-off velocity method and center of jumper's body vertical position method two different methods of numerical integration are compared in this paper.

2 Mechanical Description of the Vertical Jump

In general there are three different kinds of vertical jumps in biomechanics: counter-movement jump (CMJ), squat jump and drop jump. In this paper the CMJ is analyzed.

During the CMJ two external forces act on the jumper body: the gravitational force and the ground reaction force (it is assumed that the air resistance is negligible). These two forces are depicted in Fig. 1. The following equation of motion can be written in such a case

$$ma_y(t) = R(t) - mg, \tag{1}$$

where m is the jumper mass, $R(t)$ is the GRF and g is the gravitational acceleration.

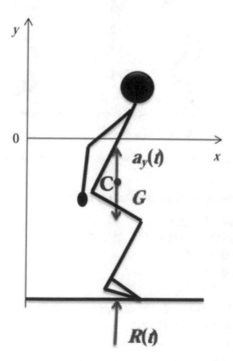

Fig. 1. External forces acting on the jumper body during the vertical countermovement jump.

Consequently the acceleration of the CJB is given by

$$a_y(t) = \frac{R(t)}{m} - g. \qquad (2)$$

The velocity and vertical position of the CJB can be obtained by integration

$$v_y(t) = v_{0y} + \int_0^t a_y(t)dt, \qquad (3)$$

$$y_C(t) = y_0 + \int_0^t v_y(t)dt, \qquad (4)$$

where v_{0y} and y_0 denote the initial velocity and vertical position of the CJB, respectively.

In vertical CMJ different phases can be distinguished. These phases are schematically depicted in Fig. 2.

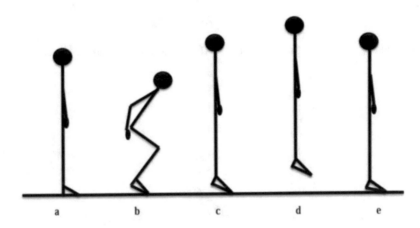

Fig. 2. Different phases of the vertical countermovement jump.

There are five characteristic phases during the CMJ:

- the stationary phase (Fig. 2a),
- the downward movement phase (Fig. 2b),
- the pushoff phase (Fig. 2c),
- the flight phase (Fig. 2d),
- the landing phase (Fig. 2e).

Figure 3 presents the GRF, velocity and position of the CJB during different phases of the vertical countermovement jump.

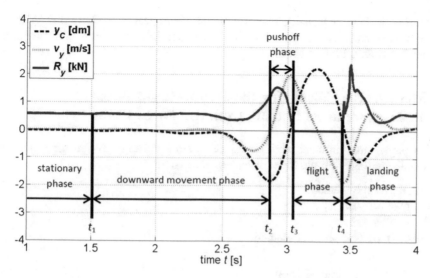

Fig. 3. Ground reaction force, velocity and position of the center of the jumper's body during the vertical countermovement jump.

In the first phase (the stationary phase) the jumper is standing upright. The GRF is constant and it is equal to the jumper's body weight. The acceleration, velocity and position of the CJB are initially equals to zero.

The downward movement phase begins when the jumper starts to move. The beginning of the phase is denoted by the instant t_1 in Fig. 3. The resultant force acting on the CJB ($R(t) - mg$) is negative at the beginning of this phase. The value of the resultant force (and the acceleration of the CJB) is decreasing. Then it achieves a minimal value and after that it is increasing. The resultant force $R(t) - mg$ is positive in the end of the phase. The velocity of the CJB is negative during this phase. In the end of the downward movement phase the velocity is equal to zero. The position of the CJB is negative and it is decreasing during the whole phase. It achieves the minimal value at the instant t_2.

During the pushoff phase the GRF is slightly increasing at the beginning of the phase. Then it achieves the maximum value of the pushoff phase and after that it is decreasing. In the end of the phase (at the instant t_3) the GRF equals to zero. The velocity of the CJB is increasing during this phase and it is slightly decreasing in the end of the pushoff phase. The position of the CJB is increasing and it a little greater than zero in the end of the pushoff phase. It means that it has the same position like initially in the stationary phase.

The GRF is equal to zero during the flight phase. It means that on the CJB acts only the gravity force. The velocity of the CJB in this phase is a linear function and it is decreasing. In the first part of the phase the velocity of the CJB is positive and in the second part it is negative. The position of the CJB is increasing in the first part of the flight phase. Then it achieves a maximal value (when the velocity of the CJB equals to zero) and in the second part of the phase it is decreasing – the jumper is falling.

During the landing phase there is an irregular sharp peak in the GRF plot. After that the GRF is equal to the jumper's body weight when there is no motion of the jumper and he is standing upright again.

3 Methods for Calculating the Flight Height Based on Measurements Coming from the Force Plates

In this paper three different methods for calculating the flight height based on measurements coming from the force plates are compared:

- the flight time method,
- the take-off velocity method,
- the work-energy method,
- the CJB vertical position method.

3.1 The Flight Time Method

In the first method two instances from the GRF curve should be read: the instant of takeoff (t_3) and the instant of landing (t_4). The difference between these two instants is the flight time

$$\Delta t_{flight} = t_4 - t_3. \tag{5}$$

In this method it is assumed that the vertical position of the CJB at the takeoff instant t_3 is equal to the vertical position of the CJB at the landing instant t_4. It results in the same times of rising and falling of the CJB during the flight phase. In such a case the maximal vertical position of the CJB is obtained exactly in the middle of the flight phase. Furthermore during this phase only the gravity force acts on the CJB. Thus the following relation can be written between the flight time Δt_{flight} and the velocity at the takeoff instant v_{t0}

$$v_{t0} = \frac{g \Delta t_{flight}}{2}. \tag{6}$$

Comparison between the sum of the gravitational potential and kinetic energies at the takeoff instant t_3 and at the instant the jumper reaches the peak of the vertical jump leads to the following relation

$$\frac{1}{2} m v_{t0}^2 + mg y_{t0} = \frac{1}{2} m v_{peak}^2 + mg y_{peak}. \tag{7}$$

Taking into account that at the instant the jumper reaches the peak of the vertical jump the velocity of the CJB equals to zero ($v_{peak} = 0$) the above equation can be rewritten in the following form

$$g\left(y_{peak} - y_{t0}\right) = \frac{1}{2} v_{t0}^2.$$ (8)

From this relation the flight height is given by

$$y_{height} = y_{peak} - y_{t0} = \frac{v_{t0}^2}{2g}.$$ (9)

Using Eq. (6) the above equation can be rewritten in the form

$$y_{height} = \frac{g \Delta t_{flight}^2}{8}.$$ (10)

3.2 The Take-Off Velocity Method

In general the vertical position of the CJB at landing instant t_4 can differ from the vertical position of the CJB at takeoff instant t_3. The jumper can fold his legs during landing in order to extend the flight time. Because of that estimating the flight height based on the flight time can generate a subjective error [12, 16–18]. In order to avoid this error the flight height can be calculated based on the velocity at the takeoff instant t_3 (read from the velocity of the CJB curve obtained from integration of the equation of motion). After that Eq. (9) can be applied directly in order to obtain the flight height y_{height}.

3.3 The Work-Energy Method

In this method the work-energy theorem is applied to the force-vertical position curve. Using this theorem to the ground contact phase (from the beginning instant t_1 to the instant of takeoff t_3) the following equation is obtained

$$\int_{y(t_1)}^{y(t_3)} (R - mg)dy = \int_{y(t_1)}^{y(t_3)} Rdy - \int_{y(t_1)}^{y(t_3)} mgdy = W_R - W_{BW} = \frac{1}{2}mv_{to}^2,$$ (11)

where W_R denotes the work done by the GRF and W_{BW} is the work done by the gravity force mg on the jumper. From the above relation the velocity at the takeoff instant v_{to} can be calculated. Based on this value the flight height y_{height} is obtained using Eq. (9). An example of the force-vertical position curve for the ground contact phase is depicted in Fig. 4.

3.4 The Center of Jumper's Body Vertical Position Method

The flight height can be also obtained directly from the vertical position of the CJB curve obtained by twice integration of the equation of motion. In order to do this only the vertical positions of the CJB at two instances should be read from the curve: at the

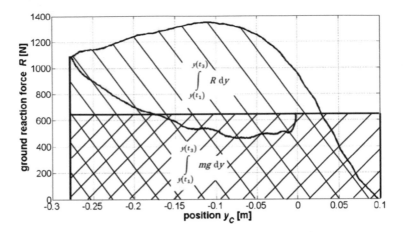

Fig. 4. The force-vertical position curve.

takeoff instant $(y(t_3))$ and at the peak position instant (y_{peak}). Then the flight height is given by

$$y_{height} = y_{peak} - y(t_3).$$ (12)

4 Results

In this study the vertical countermovement jumps were performing by two persons (person A and person B). Their physical parameters are collected in Table 2.

Table 1. Physical characteristics of persons A and B.

Parameter	Person A	Person B
Age	22	22
Height	170 cm	164 cm
Mass	65 kg	59 kg

Each person performed 100 jumps. In order to ensure repeatability the same procedure was applied during measurements. The jumper started performing the jump with the same position at the same place. Furthermore, after each jump the jumper had a rest in order to avoid muscles fatigue.

The flight height was calculated based on GRF in the paper. Any other measurements were not performed in this study because the aim of it was to compare different methods for calculating the flight height based on the GRF. All calculation were done in MATLAB software.

Table 2. Comparison of the results using the trapezoidal rule.

Method	Flight height for person A	Flight height for person B
The flight time method	23.46 ± 3.67 cm	21.83 ± 2.32 cm
The take-off velocity method	21.50 ± 3.10 cm	20.54 ± 2.31 cm
The work-energy method	23.05 ± 3.18 cm	22.27 ± 2.26 cm
The CJB vertical position method	20.98 ± 3.06 cm	20.03 ± 2.28 cm

Table 2 presents results obtained for four different methods of calculating the flight height based on the GRF. In all these calculations the trapezoidal rule was applied in order to integrate the equation of motion.

The first method is based only on the flight time. The rest of compared methods need any method of integration. In the take-off velocity method the integration process is applied once for obtaining the velocity of the CJB. In the last method (the CJB vertical position method) the integration process is used two times (by the first time for obtaining the velocity and by the second time for obtaining the position of the CJB). In the third method (the work-energy method) the integration process is applied three times – twice for obtaining velocity and position of the CJB and by the third time for obtaining the work done by the GRF. It can be notice in Table 1 that the differences between the results obtained using different methods are relatively high.

Another method for integration is applying firstly the cubic spline interpolation [19]. After this interpolation each curve is given in an analytical form which can be directly integrated without using any numerical method. The results obtained for the take-off velocity method and the CJB vertical position method using the cubic spline interpolation are shown in Table 3.

Table 3. Comparison of the results obtained using the cubic spline interpolation.

Method	Flight height for person A	Flight height for person B
The take-off velocity method	22.02 ± 3.13 cm	20.85 ± 2.93 cm
The CJB vertical position method	22.01 ± 3.13 cm	21.38 ± 3.16 cm

The differences between the results obtained for the take-off velocity method and CJB vertical position method using the cubic spline interpolation are smaller than using the trapezoidal rule.

5 Conclusions

The results presented in the paper show that not only method of estimating the flight height (based on flight time, take-off velocity, work-energy method or the CJB vertical position) has a big influence on the results obtained from the force plates measurements. In all compared methods any integration is needed. Using the cubic spline interpolation for the obtained data and then integrating these splines the differences between the flight height obtained for different methods can be smaller than using the trapezoidal rule.

Acknowledgments. The work was supported by the grant 02/21/DSPB/3493 founded by the Ministry of Science and Higher Education, Poland. During the realization of this work Dr. Jakub K. Grabski was supported with scholarship funded by the Foundation for Polish Science (FNP).

References

1. Payton, C.J., Bartlett, R.M. (eds.): Biomechanical Evaluation of Movement in Sport and Exercise. Routledge, London (2008)
2. Corazza, S., Mündermann, L., Chaudhari, A.M., Demattio, T., Cobelli, C., Andriacchi, T.P.: A markerless motion capture system to study musculoskeletal biomechanics: visual hull and simulated annealing approach. Ann. Biomed. Eng. **34**(6), 1019–1029 (2006)
3. Walczak, T., Grabski, J.K., Grajewska, M., Michałowska, M.: Application of artificial neural networks in man's gait recognition. In: Kleiber, M., et al. (eds.) Advances in Mechanics: Theoretical, Computational and Interdisciplinary Issues. Proceedings of the 3rd Polish Congress of Mechanics (PCM) and 21st International Conference on Computer Methods in Mechanics (CMM), pp. 591–594. CRC Press, Taylor & Francis Group, London (2016)
4. Cross, R.: Standing, walking, running and jumping on a force plate. Am. J. Phys. **67**(4), 304–309 (1999)
5. Grabski, J.K., Walczak, T., Michałowska, M., Cieślak, M.: Gender recognition using artificial neural networks and data coming from force plates. In: Gzik, M., et al. (eds.) Innovations in Biomedical Engineering, IBE 2017. Advances in Intelligent Systems and Computing, vol. 623, pp. 53–60. Springer, Cham (2018)
6. Walczak, T., Grabski, J.K., Grajewska, M., Michałowska, M.: The recognition of human by the dynamic determinants of the gait with use of ANN. In: Awrejcewicz, J. (ed.) Dynamical Systems: Modelling. Springer Proceedings in Mathematics and Statistics, vol. 181, pp. 375–385. Springer, Cham (2016)
7. Merletti, R., Parker, P. (eds.): Electromyography. Physiology, Engineering and Noninvasive Applications. John Wiley & Sons, Hoboken (2004)
8. Grabski, J.K., Kazimierczuk, S., Walczak, T.: Analysis of the electromyographic signal during rehabilitation exercises of the knee joint. Vibr. Phys. Syst. **26**, 79–86 (2014)
9. Nathan, R.H.: A dynamometer for biomechanical use. J. Biomed. Eng. **1**(2), 83–88 (1979)
10. Grygorowicz, M., Michałowska, M., Walczak, T., Owen, A., Grabski, J.K., Pyda, A., Piontek, T., Kotwicki, T.: Discussion about different cut-off values of conventional hamstring-to-quadriceps ratio used in hamstring injury prediction among professional male football players. PLoS ONE **12**(2), e0188974 (2017)
11. Maund, P.J., Foster, C.: Physiological Assessment of Human Fitness, 2nd edn. Human Kinetics (2006)
12. Monnet, T., Decatoire, A., Lacouture, P.: Comparison of algorithms to determine jump height and flight time from body mounted accelerometers. Sports Eng. **17**(4), 249–259 (2014)
13. Casartelli, N., Muller, R., Maffiuletti, N.A.: Validity and reliability of the myotest accelerometric system for the assessment of vertical jump height. J. Strength Cond. Res. **24**(11), 3186–3193 (2010)
14. Nuzzo, J.L., Anning, J.H., Scharfenberg, J.M.: The reliability of three devices used for measuring vertical jump height. J. Strength Cond. Res. **25**(9), 2580–2590 (2011)

15. Carlos-Vivas, J., Martin-Martinez, J.P., Hernandez-Mocholi, M.A., Perez-Gomez, J.: Validation of the iPhone app using the force platform to estimate vertical jump height. J. Sports Med. Phys. Fit. **58**(3), 227–232 (2018)
16. Castagna, C., Ganzetti, M., Ditroilo, M., Giovannelli, M., Rocchetti, A., Manzi, V.: Concurrent validity of vertical jump performance assessment systems. J. Strength Cond. Res. **27**(3), 761–768 (2013)
17. Linthorne, N.P.: Analysis of standing vertical jumps using a force platform. Am. J. Phys. **69**(11), 1198–1204 (2001)
18. Musayev, E.: Optoelectronic vertical jump height measuring method and device. Measurement **39**(4), 312–319 (2006)
19. Bartels, R.H., Bealty, J.C., Barsky, B.A.: Hermite and Cubic Spline Interpolation. Morgan Kaufmann, San Francisco (1998)

High Efficient Weightlifting Barbell Tracking Algorithm Based on Diamond Search Strategy

Ching-Ting Hsu$^{(\boxtimes)}$, Wei-Hau Ho, and Jen-Shi Chen

Graduate Institute of Sports Equipment Technology,
University of Taipei, Taipei, Taiwan, R.O.C
jingting@utaipei.edu.tw

Abstract. An efficient weightlifting barbell tracking algorithm has been proposed in this paper. We aim to fast and accurately extract barbell route from weightlifting competition video sequence for training. To achieve this target, a vertical enhancement diamond search pattern is adopted to find out the most similarity areas. From the experimental result, our proposed algorithm is able to keep tracking exactness of barbell object and respond in real time. It helps athletics, coaches and biomechanics scholars to gather the weightlifting performance as rapidly as required by the user.

Keywords: Weightlifting · Object tracking · Computer Aided Sport Training

1 Introduction

Weightlifting is one of the focuses in Olympic training programs, especially the female weightlifting in Taiwan. Taiwan national female weightlifting team has brilliant achievements. Among them, Hsu Shu-Ching won a female 53 kg silver medal in 2012 London Olympic Games, which attracted people's attention to weightlifting. She even won the gold medal and broke the world record in 2014 Incheon Asian Games that made biomechanics researchers get interested in doing research in weightlifting. Additionally, Kuo Hsing-Chun, female 58-kg class player claimed the gold medal, and broke the clean&jerk world record at 2017 Taipei Summer Universiade. They are considered the pride of Taiwan.

Weightlifting coaches often use mirrors in the training ground. While training, mirrors are very usefully for athletes to know whether their gestures are correct or not. The mirror response the action in real time, but sometimes the athletes may forget their gesture in a short while after finish the weightlifting. Hence, the effectiveness of training is decreased. To solve this issue, camcorders are also used. Mirrors and camcorders, these auxiliary training methods or equipments are used to let athletes to observe their actions. However, if coaches don't have enough experience or the proper method to find the tiny differences, they may find difficulty in observing improvement or physical fatigue [1].

As one of the research topics or training methods of sport biomechanics, kinematic analysis has been widely researched and acceptable applied for sport performance evaluation. Kinematic analysis software gathers important kinematic parameters such as

K. Arkusz et al. (Eds.): BIOMECHANICS 2018, AISC 831, pp. 252–262, 2019.
https://doi.org/10.1007/978-3-319-97286-2_23

trajectory or moving speed, which help coaches, athletes or researchers to understand the performance of the athletes. Indeed, manually operation of the professional kinematic analysis software consumes huge operation time, and difficultly usage may also reduce the usage intention of the coaches or athletes. Sport video contains a lot of useful information. Not just for training course, they are very important in the competition, too. Currently, recoded sport video is one of the most effective schemes, which has accessible and convenient features. Coaches give the players their feedback of the kinematic analysis by recoding motion pictures. Not only for self-observation and evaluation, the sport video can further gather the rival's intelligence. This intelligence helps us to understand rival's strategy and then develop new tactic to win the competition.

Basically, professional kinematic software such as Siliconcoach [2], SIMI [3] or other related software are very popular to analyze kinematic. This kind of the software can provide some kinematic data analysis, for example, the barbell trajectory, the overall time and the angle and limb of each phase. However, the common problems of this kind of the software are to waste long time to complete a full kinematic analysis, to operate manually, and too complex to understand intuitively. Another point of view, users must have enough training if they need to operate, read and further understand the meaning of the kinematic parameters. In general, weightlifting is a sport that takes about 30 s to complete. In other words, to complete and analysis video sequences of 30 s, it usually takes more then 10 min. Unfortunately, long analysis time, the difficult usage and costly training, all of them exhausted the coaches' and athletes' willingness of scientific sport training. Besides, it does not achieve immediate information feedback and practical requirements.

Video analysis is a popular research topic in computer vision. To analyze sport videos for training purpose or for strategy discussion. Traditional kinematic analysis software adopts the advent of 3D opto-reflective marker based require expertise for body marker placement. These opto-reflective markers help motion analysis software to create human body model and track the limb moving in high accuracy. But user may spend huge time to setup the experimental environment. Furthermore, most of the kinematic analysis software operating with infrared may not suit for outdoor using. Although this analysis equipment and scheme may provide high accuracy analysis result, however, these limitations restrict the growth of the computer aided sport training. On the other hand, these markers restrict the body moving and further decrease the training performance.

Trajectory is an important indicator of kinematic analysis parameters, it is able evaluate the effectiveness of athlete training. In past year, many researchers are interested in the barbell moving. Sato et al., Lenjannejadian et al. and Rahmati et al. utilized sensors such as triaxial accelerometer to gather the trajectories of the barbell [1, 4, 5]. Sensor is one of the effective and popular ways to gather the barbell moving path. However, when the weightlifter puts down the barbell after finishing the action, the total weight is nearly 170.1G [1]. This will affect the accuracy and even damage sensors. Although sensors can provide kinematic parameters in high accuracy, however, they cannot be suitable in the real competition. It means that we cannot use sensor to obtain athletes' real data in a competition.

Taking all these factors into consideration, pure video analysis technique is the best way to gather the barbell trajectory. It is able to apply to daily weightlifting training and really competitions. As we've mentioned, traditional video kinematic analysis software uses huge operation time. Hence, our aim is to propose an efficient barbell tracking algorithm to extract the trajectory from the weightlifting video sequence without any markers and sensors.

In this paper, the characteristics of moving barbell have also been added to our design, the diamond search algorithm [6] is modified and improved to find the selected barbell in the video sequence. Pixel value in barbell area as the feature vector and will be searched in further frames in the video sequence by diamond search pattern. We take the video sequences which captured from 2017 Taipei Summer Universiade as the testing ground. The proposed algorithm and results will be described later.

The rest of this paper is organized as follows. Section 2, relevant works are reviewed. The proposed modified diamond search algorithm will be introduced in Sect. 3. Experimental results and conclusions are discussed and shown in Sects. 4 and 5, respectively.

2   Relevant Works

Enoka et al. [7], Garhammer et al. [8] pointed out that weightlifting athletes need technique, power and flexibility skill. Weightlifting is a continuous movement which lifter pulls the barbell from the ground and lifts the arms straight to top of the head. Storey et al. took snatch as an example and they pointed out that the snatch movement from start to end took approximately 3 to 5 s [9]. This observation is consistent with our experience. Decomposing complete actions, including the athlete goes up to the platform, finishes the action and finally leaves the platform, all of them take about 10–30 s. Because weightlifting is a short time cycle sport, Sato et al. pointed out that simple system is necessary of providing real-time information for athletes and coaches and that can help training and competition analysis [1]. Refer to International Weightlifting Federation technical and Competition rules and regulations [9], the kinematic analysis results should be provided within 90 s for satisfying in a real application.

Barbell kinematic parameters are common observation tool for sport performance in weightlifting researches, especially the barbell trajectory, velocity, acceleration, and combined with joint angles [1]. Harbili denoted that barbell largest deviation in the vertical axis is horizontal displacement. Harbili indicated that the largest deviation of barbell at the vertical axis is horizontal displacement. When the trajectory at vertical axis deviation is too large, the barbell is probably to be in an unstable state, this may cause athlete does not fluent when lifting the further to no lift [11]. Sensor is a way to obtain the characteristics of the barbell movement in space [1, 4, 5], acceleration sensors are utilized to gather the three-dimensional movement of the barbell in position. After projecting the three-dimension data to two-dimensional coordinate, the barbell moving trajectory is established. Through sensor, either acceleration or movement distance of objects, can be accurately use. However, sensors setup on the top of the barbell, the barbell self-turning may cause sensor misjudgment or the psychological impacts of the athletes are in need of special consideration [12]. Even, Sato et al. [1] described that when weightlifters complete the lifting, the barbell drop down from the air will produce 170.1G of gravity. The sensor may be misalignment and even damaged, also, the electronic sensors are not suitable in official competitions. Researches cannot utilize the sensor to observe lifters' real athletic performance during the competition. To aid weightlifting training, barbell trajectory obtained from the sport video by computer is required with the high efficiency and accuracy [12].

Ren et al. obtained the characteristics of the weightlifting barbell and dumbbell motion in the video sequences, which are periodical activities. Hence, they utilized the periodical motion pattern to detect and count the barbell activity [13]. Low-level features such like color histogram are also considered. Jocic et al. utilized illumination change and OpenGL library [14]. Zikovic et al. proposed 5-degree of freedom color histogram based non-rigid object tracking algorithm combined with mean-shift method [15]. Color information is easy to be gathered, however, the color information is light sensitive which may be influenced by light changing to further decrease the recognition and tracking accuracy.

Above object tracking methods gather the trajectory of the video object by considering the low-level features with a kernel function. However, both of the feature extraction procedure and similarity calculation consume huge computational power. On the other hand, most of the object tracking algorithm is based on training samples of the target object. Accordingly, in our proposed barbell tracking algorithm, we gather the local feature without any sampling procedure. After gather the local feature, we compare the similarity directly with a fast motion estimation scheme. That achieves both efficiency and accuracy with low computational complexity.

3 Proposed Efficiency Weightlifting Barbell Tracking Algorithm

Our proposed efficient weightlifting barbell tracking algorithm is shown in this section. Our aim is to create a high efficiency and high accuracy computer aided weightlifting system with barbell trajectory. The flowchart and detail of the algorithm are shown as follows.

3.1 Flowchart

Figure 1 shows the flowchart of our proposed algorithm. For wide range using, we consider general RGB video sequence since the user can take the videos from a training course or competition even from the video website such like YouTube [16]. The user should select the barbell in first frame as the sampling area and this area will be saved to the buffer for further searching in next frame. From the second frame, the sampling area from previous frame utilizes our proposed Vertical Enhancement Diamond Search Pattern to find out the most similarity area in current frame. The most similarity area in current frame must be the target weightlifting barbell we wanted. The barbell movement from previous frame to current frame denotes as a motion vector with vertical and horizontal components. Then sampling area will be updated by weightlifting barbell in current frame, and the motion vector will be saved.

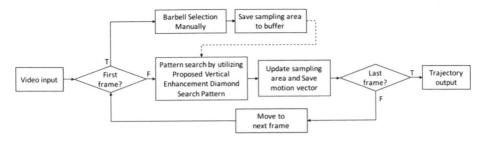

Fig. 1. Flowchart of the proposed efficient weightlifting barbell tracking algorithm

After pattern searching finished, our proposed algorithm will check if the current frame is last frame. If not, the sampling area will be updated and continue searching in next frame. If the current frame is last frame, motion vectors in each frame will be connected as the barbell trajectory.

3.2 Pattern Search by Proposed Vertical Enhancement Diamond Search Pattern

Motion estimation and motion compensation are two key features in video coding system to effectively reduce the data redundancy. Almost all video codecs utilize block-base motion estimation scheme which divide coding frame into $m \times m$ size macroblock and then utilize these macroblocks to compare the similarity between current frame and reference frame. Figure 2 shows an example of block-base motion estimation procedure.

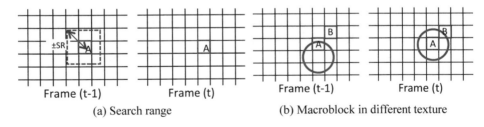

(a) Search range

(b) Macroblock in different texture

Fig. 2. Example of block-base motion estimation

In Fig. 2(a), we can see a macroblock named A in current frame, *Frame(t)*, and co-located position in reference frame, *Frame(t − 1)*. Motion estimation scheme takes this macroblock in *Frame(t)* and compare the similarity in co-located position with a searching range, ±*SR*. The most popular similarity function is sum of absolute difference (SAD) as Eq. (1).

$$SAD(x,y) = \sum_{i=0}^{m-1} \sum_{j=0}^{m-1} |S(x+i, y+j) - T(x+i, y+j)| \qquad (1)$$

We consider a condition that an object covers to many macroblocks shown in Fig. 2 (b). In Fig. 2 (b), although macroblock A and B are covered by green circle in *Frame(t)*, but the texture in these two macroblocks are different. On the other hand, macroblock A and B in *Frame(t)* are covered by green circle, however, the co-located A and B in *Frame(t)* are not. Above condition may cause different motion vector direction in one object and further affect the accuracy of the object tracking algorithm. Hence, in this paper, we consider to compare the similarity by object to instead of block-base searching scheme. After user selected the target barbell, our proposed algorithm tracks the barbell moving automatically. Figure 3 shows examples of the sampled barbell by user selection from the first frame in different video sequences.

Fig. 3. Barbells sampled by user from the first frame of the video sequence

After target barbell selected, we then utilized our proposed vertical enhancement diamond search pattern to find the most similarity area in next frame. Diamond search algorithm was proposed by Zhu et al. in 2000 [17]. They use two different patterns, large diamond search pattern and small diamond search pattern, to obtain the most similarity area in reference frame. Considering the moving characteristics of the barbell which is almost moving on vertical axis, we modify the diamond search algorithm to enhance the searching performance on vertical axis.

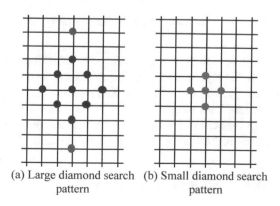

(a) Large diamond search   (b) Small diamond search
         pattern   pattern

Fig. 4. Proposed vertical enhancement diamond search pattern

Figure 4 shows our proposed vertical enhancement diamond search pattern. We extend large search pattern on vertical axis as the blue point shown in Fig. 4(a). Similar to original diamond search scheme, we use large diamond search pattern in advanced. Points in Fig. 4 denote the coordinate in next frame and we also use SAD to evaluate the similarity between sampled barbell and the candidate barbell on these coordinates. If the minimum SAD value does not obtain on the center point, the large diamond search pattern will then be moved to the point with minimum SAD value. We have to note that the large pattern moving almost converges in three time of the movement. Figure 5 shows movement examples of our proposed vertical enhancement diamond search pattern. As mentioned above, if the minimum SAD position obtained on vertical or horizontal axils (Fig. 5(a)) or minimum SAD point obtained on corner points (Fig. 5 (b)), the large diamond search pattern will be moved to the minimum SAD point. While the large diamond search pattern does not update (in center), then the small diamond search pattern (Fig. 4 (b)) will then be utilized to refine the motion vector (as can be seen in Fig. 5(c)).

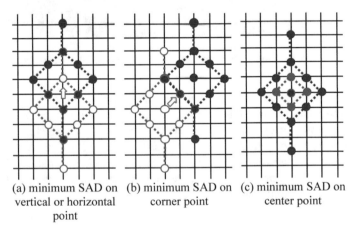

(a) minimum SAD on   (b) minimum SAD on   (c) minimum SAD on
vertical or horizontal   corner point   center point
point

Fig. 5. Pattern movement examples of vertical enhancement diamond search algorithm

Since the small diamond search pattern is utilized to refine the motion vector, it will not be moved anymore. Our proposed algorithm finished searching when the small diamond search pattern gathers the minimum SAD value. The point with minimum SAD value will be the target barbell position in current frame. Thus, the coordinate of the new position will be updated as the motion vector and sampled area will be updated too. The motion vector will further be connected into the barbell's trajectory.

4   Experimental Results and Discussion

The experimental results are shown and discussed in this section. Our testing videos are captured from 2017 Taipei Summer Univresiade. We use women 63 kg Group A snatch competition as our testing bed. There are totally 11 lifters with 35 attempt videos. These videos are captured by a consumer digital video camcorder with full HD resolution (1920 × 1080 and 60 frames per second) and H.264 video bitstraem. The proposed algorithm is implemented in MacOS 10.12.6 Sierra with QT and openCV library [18].

The developing platform is Mac mini with 2.6 GHz Intel Core i5 processor and 8 GB RAM. Not only implement our proposed vertical enhancement diamond search algorithm, we further realize full search algorithm as our comparison target. For consistent with all video sequence, we set the sampling barbell size is 40 × 40 pixels and search range of full search is ±32 pixels.

We show the performance of our proposed algorithm in advanced. Our proposed algorithm average consumes 2.4 ms for searching the similarity barbell in each frame. Comparing to full search which costs 203 ms, the tracking efficiency has a significant improvement. Furthermore, our proposed algorithm has approximating 93% the same with full search's result. Averaging 7% miss-matching condition will be discussed later.

(a) Full search (SR: ±32)   (b) Proposed algorithm

Fig. 6. Subjective quality comparison of full search and proposed algorithm

The subject quality is shown in Fig. 6. In this figure, (a) and (b) are the result of full search and proposed vertical enhancement diamond search algorithm, respectively. The trajectory and markers are draw on-the-fly. In Fig. 6, green solid line is the trajectory which means the barbell moving route from barbell rising up and droping down. Yellow circles are the markers which depict the position of the barbell in each frame we projected to the current frame. Users can use the trajectory and distance between position marker to observe and evaluate the lifter's performance. From Fig. 6, the trajectories in (a) and (b) are very similarity. Furthermore, from the position marker, we provide a visualization tool to observe the sport performance. For example, the distance between any two yellow circles can be shown as the barbell movement duration between these two frames. It means that users can obtain the moving speed of the barbell though these markers.

Figure 7 shows the same lifter in different attempt. There are 83 kg on the bar at first and second attempt, and first attempt is no lift. From Fig. 7, we can observe that the trajectory of the no lift is excursive. Not only compare their self, Fig. 8 compares to

(a) 1st attempt, No lift   (b) 2nd attempt, good lift

Fig. 7. Compare with different attempts

(a) Trajectory of golden metal   (b) Another lift (good lift)
(good lift)

Fig. 8. Compare with different lifters

two different lifters. Since Fig. 8 (a) is the golden metal winner, users can observe their barbell trajectory to obtain the difference between two lifters' skill. Although Figs. 7 and 8 show a simple and easy functions, however, the significant improvement is that the user can get this high accuracy comparison results in one minute.

Finally, we discuss about the miss-match problem. As mentioned, we enhance the vertical direction search pattern but not horizontal because the considering of barbell moving characteristics. Sometimes if the barbell moving on the horizontal in high motion condition, our algorithm may not be suit. Figure 9 shows an example of the barbell miss-match. In this case, after catching the bar, the lifter tries to control the barbell to prevent the barbell full down. This huge horizontal moving causes our search pattern exception. This situation can be considered in the future work.

     (a) Full search   (b) Proposed algorithm

Fig. 9. Miss-match condition

5 Conclusions

An efficiency weightlifting barbell tracking algorithm is proposed in this paper. We consider the moving characteristics of the barbell to design and implement a vertical enhancement diamond search pattern. The vertical enhancement diamond search scheme significantly decreases the computing time and achieve the tracking performance.

From the results, we use general RGB video captured in a real-competition, 2017 Taipei Summer Universiade to prove the quality and performance. Comparing to full search, our proposed algorithm archives similar quality with significant lower computational power. We also show the application of our proposed algorithm that we compare any two different trajectories efficiently. Trajectory comparison for two different lifters or two different attempts can be provided in a very short time. Comparing to traditional kinematic analysis software needs more than 10 min, our proposed vertical enhancement diamond search pattern makes computer aided weightlifting training more easily.

6 Acknowledgement

This research is supported by Ministry of Science and Technology, Taiwan, R.O.C. The project number is 106-2410-H-845-018.

References

1. Sato, K., Sands, W.A., Stone, M.H.: The reliability of accelerometer to measure weightlifting performance. Sports Biomechanics **11**(4), 524–541 (2012)
2. Siliconcoach official website. https://www.siliconcoach.com/products/pro8. Accessed 25 Mar 2018
3. SIMI official website. http://www.simi.com/. Accessed 25 Mar 2018
4. Lenjannejadian, S., Rostami. M.: Optimal trajectories of snatch weightlifting for two different weight classes by using genetic algorithm. In: 2008 Cairo international conference on biomedical engineering, pp. 1–4, Cairo, Egypt (2008)

5. Rahmati, S.M.A., Mallakzadeh, M.: Determination of the optimum objective function for evaluation optimal body and barbell trajectories of snatch weightlifting via genetic algorithm optimization. In: 18th Iranian Conference on Biomedical Engineering, Iranian (2011)

6. Zhu, S., Ma, K.K.: A new diamond search algorithm for fast blocking-matching motion estimation. IEEE Trans. Image Process. **9**(12), 287–290 (2000)

7. Enoka, R.M.: The pull in Olympic weightlifting. Med. Sci. Sports **11**, 131–137 (1979)

8. Garhammer, J.: Weight lifting and training. Biomechanics of sport, pp. 169–211 (1989)

9. Storey, A., Smith, H.K.: Unique aspects of competitive weightlifting. Sports Med. **42**(9), 769–790 (2012)

10. Aján, T., et al.: 2018 International weightlifting federation technical and competition rules and regulations. In: International Weightlifting Federation (2018)

11. Harbili, E.: A gender-based kinematic and kinetic analysis of the snatch lift in elite weightlifters **11**(4), 162–169 (2012)

12. Hsu, C.T., Ho, W.H., Chen, J.L., Lin, Y.C.: Efficient barbell trajectory extraction algorithm for kinematic analysis using video spatial and temporal information. In: 2014 International conference on biomedical engineering. Zurich, Switzerland (2014)

13. Ren, Y., et al.: An efficient framework for analyzing periodical activities in sports video. In: 4th International Conference on Image and Signal Processing, pp. 1–50

14. Jocic, M., Oradovic, D., Kojovic, Z., Tertei, D.: OpenGL implementation of a color based object tracking. In: 3rd International Conference on Information Society Technology, pp. 7–11. Toronto, Canada (2013)

15. Zivkovic, A., Krose, B.: An em-like algorithm for color-histogram-based object tracking. In: 2004 IEEE Computer Society Conference on Computer Vision and Pattern Recognition, I-798–803. Washington DC, USA (2004)

16. YouTube official website: https://www.youtube.com/. Accessed 29 Mar 2018

17. Zhu, S., Ma, K.K.: A new diamond search algorithm for fast block-matching motion estimation. IEEE Trans. Image Process. **9**(2), 287–290 (2000)

18. OpenCv official website: https://opencv.org/. Accessed 30 Mar 2018

Are Leg Electromyogram Profiles Symmetrical During Full Squat?

Henryk Król$^{(\boxtimes)}$ and Krzysztof Kmiecik$^{(\boxtimes)}$

The Jerzy Kukuczka Academy of Physical Education in Katowice,
Katowice, Poland
h.krol@awf.katowice.pl, krzysiek-kmiecikl@o2.pl

Abstract. In order to see how electromyogram (EMG) profiles change during the squat movement with increasing loads, we determined the degree of symmetry of selected homologous muscles. Seven healthy men (age range 20–42 years), recreationally performing strength exercises voluntarily participated in the research. The participants varied in height from 172 to 183 cm and in mass from 74 to 94 kg. The participants performed consecutive sets of a single repetition of full back squatting, each time with an increased load (70–100% 1RM). To record the parameters of participant and barbell movement, the Smart-E measuring system (six infrared cameras and a wireless module for measuring the bioelectric activity of muscles) were used. Electrical activity was recorded using surface electrodes for muscles on both sides of the body (homologous): *tibialis anterior* (TA), *gastrocnemius medialis* (G$_{med}$), *biceps femoris* (BF), *rectus femoris* (RF), *gluteus maximus* (G$_{max}$) and *erector spinae* (ES). The mean of averages of modules of amplitude differences (MAMAD) between individual pairs of normalized homologous muscles for squat movement was accepted as a measure of symmetry of homologous muscles. A statistically significant increase of symmetry of EMG profiles (MAMAD) with increasing loads was seen only for TA and ES muscles and only in two of the six of analyzed cases. Of particular interest was the statistically significant large MAMAD increase for BF and ES muscles and for the so called prime mover (G$_{max}$) in the ascent phase of the squat compared to the descent phase.

Keywords: Barbell squat · Electromyography · Muscle symmetry

1 Introduction

Individuals performing full squats with a maximum load often show altered movement patterns compared to trials with moderate loads. Such altered patterns of movement have also been demonstrated, e.g., in individuals with anterior cruciate ligament (ACL) injury while gait, performing functional movements and common rehabilitative exercises [7, 16, 17]. This may be partly due to impaired sensorimotor control. The action of the neuromuscular system that control a movement reflects the electromyographic (EMG) profiles, in our case, muscles of the lower limbs. Thus, it seems that there should be some type of relationship between EMG profiles and motor patterns.

© Springer Nature Switzerland AG 2019
K. Arkusz et al. (Eds.): BIOMECHANICS 2018, AISC 831, pp. 263–275, 2019.
https://doi.org/10.1007/978-3-319-97286-2_24

According to Yang and Winter [21], "movement is the result of the interaction between muscle tension, dictated by the central nervous system, and the mechanical demands of the task." However, understanding how the nervous system responds to changes in the mechanical demands of a task is essential to understanding motor control. The mechanical requirements of a squat with increasing loads vary considerably. For this reason, squatting with increasing loads can be effectively used to investigate the relationship between these demands and the behavior of the nervous system.

As is well known, the EMG (the electrical activity of muscle contraction) reflects both the output of the nervous system and the input to the mechanical system. Hence, the EMG amplitudes and their course over time can provide information about both of these systems. Changes in amplitude values during a barbell squat have been studied by Gullett et al. [5], and in hip thrust exercises by Contreras et al. [3]. They presented only the results of a global measure i.e. averaged EMG for the whole movement or for the individual phases (descent and ascent). Such a global measure could mask important differences in the time course of the EMG during a squat with a barbell. Therefore, there is a real need to determine EMG changes more accurately in squats with different loads, according to both amplitude and time.

The contralateral limb can potentially serve as a valuable control because one can avoid population variables as well as other variables (e.g. squatting speed). However, to ensure that the given profiles of homologous muscles are in fact different, we must quantitatively determine their degree of symmetry.

Robertson et al. [15] presented the profiles of muscle activity in the whole squat movement as linear envelope electromyograms, normalized to each subject's maximum voluntary contraction (MVCs). However, he did not investigate symmetry. In another paper, presenting mean EMG activity of seven selected muscles at regular intervals [22], similarly did not study symmetry. In a study on walking adults, Arsenault et al. [1] paid particular attention to symmetry when there was a high correlation between the shape of whole waveforms of EMG signals for homologous rectus femoris and soleus muscles. Thus, there is a need to determine more precisely the EMG changes, i.e., the degree of symmetry of homologous muscles, with increased squatting loads according to both amplitude and time. The purpose of this study was to determine the ensemble average EMG profiles in six pairs of homologous leg and hip muscles associated with increased squat loads to determine the statistical degree of their symmetry.

2   Methods

2.1   Participants

To provide a better understanding of the interplay between alterations in sensorimotor control and biomechanical joint stability following varying loads, and to further explore full squats used in resistance training, we investigated muscle activity profiles in 11 individuals with four loads (light, moderate, sub-maximum and maximum). However, for technical reasons we present only the results of seven male volunteers (age range 20–42 years, average age 27 years). The participants varied in height from 172 to 183 cm and in mass from 74 to 94 kg. All participants were experienced in performing squats (5.3 ± 0.71 years) and had no history of orthopedic injury or surgery that would have limited their ability to perform the squatting technique. The mean 1RM (one repetition maximum; [2]) load that were employed during testing were 135.7 ± 44.3 kg, and this was 160.3 ± 43.7% of their body mass. Before participation, informed consent was obtained from each volunteer. The research project was approved by the Committee of Bioethics of the Jerzy Kukuczka Academy of Physical Education in Katowice, Poland.

2.2   Research Protocol and Procedures

All participants were tested under the same conditions, in a laboratory setting. The protocol included a full back squat with free weights and the "touch-and-go" technique. The research information for the study was collected during the warm up and the main (measurement) session. Before starting the measurement session, the subjects were asked to warm up with their own routine for typical training. After a general warm-up, all participants performed a more specific warm-up that consisted of three sets of 5 to 3 repetitions with light weights selected by the subjects (at 40 ÷ 60% 1RM of the full back squat). In the main session, the participants performed consecutive sets of a single repetition of full back squatting, and each time with increased loads (70, 80, 90, and 100% 1RM the anticipated maximum weight), until the appointment of one repetition maximum. When a participant reached the anticipated maximum weight, the load was increased until the participant could no longer perform a correct full back squat. Registered attempts that constituted approximately 70, 80, 90, and 100% of 1RM were chosen for analysis (Table 1).

Table 1. Attempts of participants (N = 7) chosen for analysis.

Variable	Mean ± SD	Range
Back squat 70% 1RM (kg)	99.3 ± 36.1	60–160
Back squat 80% 1RM (kg)	112.1 ± 39.8	70–180
Back squat 90% 1RM (kg)	125.0 ± 43.5	80–200
Back squat 100% 1RM (kg)	135.7 ± 44.3	90–210

If the previous loads did not include these values, the participant would perform a full squat with the missing load after the maximum attempt. In total, each participant performed between five and eight attempts in the main session. The rest periods between the trials lasted about 4 min and were provided to avoid muscular fatigue. Exercise was begun with a given verbal command. All participants in each trial were asked to make a full squat. A detailed description of the full back squat is given in Delavier's book [4]. Two research assistants acted as spotters standing near the bar in the event that the participants were not able to successfully lift the weight. The spotters sometimes assisted the man in lifting the barbell from a support rack but the weight lifter was not assisted by the spotters during the lift.

2.3   Instrumentation and Data Collection

Movement analysis was made with the measuring system Smart-E (BTS Company, Milan, Italy). The system consisted of six infrared cameras (120 Hz) and a wireless module to measure muscle bioelectric activity (Pocket EMG). Thus, the kinematic and electromyographic data were collected simultaneously.

Three-Dimensional Kinematics. A set of passive markers reflected the infrared radiation (IR), permitting the calculation of some chosen motion parameters of the barbell, and the subject's legs. Suitable spatial accuracy was achieved by attaching the passive test markers in the axis of rotation of the lower limb joints of the subjects, and in the center and at the ends of the barbell. Based on the deterministic registered trajectory of the barbell, after its smoothing, the whole squat movement was divided into descent and ascent phases. Smart software (Smart Capture, Smart Tracker and Smart Analyzer; BTS) was used for making 3D models and for calculating the parameters. However, for technical reasons in quantitative analysis of back squats we used only the coordinate in the sagittal plane.

Electromyography. The electromyography signals were monitored and recorded using H124SG disposable electrodes. Two surface electrodes were placed 2 cm apart over the motor activation points of the tibialis anterior (TA), gastrocnemius mediale (Gmed), rectus femoris (RF), biceps femoris (BF), gluteus maximus (Gmax), and erector spinae (ES), in accordance with European Recommendations for Surface Electromyography – SENIAM [6] and secured with athletic tape. All electrodes were placed bilaterally on the leg and hip muscles. Before electrode placement, the skin surfaces were cleansed with alcohol and shaved when necessary. All electrodes remained in place until the end of all trials. Cables from the electrodes to the transmitter were secured to the participant with athletic tape to minimize distraction of the subject and interference to the EMG signal. The transmitter was placed in a belt pack worn snugly around the subject's waist. EMG signals were taken at 1 kHz sample rate. All active channels had the same measuring range and were fitted to the subject (typically ± 5 mV). Analog signals were converted to digital with 16 bit sampling resolution and collected on the measuring unit. The signals were transmitted immediately after a single trial to a computer via a Wi-Fi network.

2.4   Electromyography Data Reduction and Testing Procedures

The raw EMG signal was filtered (passband Butterworth filter, 10–250 Hz). Next, the full-wave was rectified and smoothed using the root mean square method with 100 ms mobile window. The root mean square of the electromyogram amplitude (RMS EMG) for all muscles was calculated for the entire range of motion of each lift, both descent and ascent phases separately. This was done for each squat under each of the four testing conditions (70, 80, 90 and 100% 1RM load). For this purpose, we used the previously mentioned Smart Analyzer software (BTS Bioengineering, Italy).

To compare muscle activity between the participants, during pretesting, maximal normalization contractions were performed for each muscle, so-called Maximum Voluntary Isometric Contraction (MVIC) normalization. This required the participant to contract each muscle against manual resistance provided by the experimenter for a maximum of 3 s. MVIC positions for the individual muscle were chosen based on Konrad's [9] proposals. The highest activity levels in the EMG during a 100 ms interval achieved in this test reflected the peak EMG of the muscles under isometric conditions. The maximal peak muscle activity was calculated and recorded from a suitable maximum contraction and all subsequent muscle activity was expressed as a percentage of this MVIC peak. MVIC data were processed in the same way as myoelectric data from the squats. For every muscle and for each of the testing conditions the linear envelope EMG signals were normalized for time and maximum EMG values. First, to determine the degree of symmetry of homologous muscles, the modulus (absolute value) of the differences of normalized EMG values between the pair of homologous muscles in 100 temporarily normalized points was calculated. These data were then averaged and finally the mean and standard deviation (SD) for all seven subjects were calculated for the entire squat movement and for each individual phase.

Thus, the degree of symmetry of EMG profiles was established using two measures: the mean of averages of modules of amplitude differences (MAMAD) between individual pairs of homologous muscles for the whole squat movement and the mean for two phases of movement calculated in the same way (descent - MAMADdes and ascent - MAMADasc, respectively). A larger value of these measures indicates lower symmetry of the EMG profiles of the muscles being compared, and vice versa, a smaller value indicates greater symmetry.

2.5 Statistical Analysis

Initially, basic methods of descriptive statistics were applied to data obtained from individual measurements. All data are recorded as mean $\pm$ standard deviation (SD). The normality of distribution estimated by the Kolmogorov-Smirnov test was not acceptable for all variables. Furthermore, for technical reasons it was not possible to register correct EMG signals for all pairs of homologous leg muscles during squatting with different loads. Therefore, to evaluate the symmetry of all EMG profiles for all participants and for all pairs of homologous muscles under the four testing conditions, we used the Wilcoxon Signed-Ranks Test. Thus, the Wilcoxon test was applied for the analysis of accepted measures of muscle symmetry (MAMAD). Differences $p \leq 0.05$ were considered statistically significant. All calculations and statistical analyses were performed using Statistica 10.1 (StatSoft, Inc.), and Microsoft Office – Excel 2010 packets.

3 Results

Due to the great similarity of the EMG profile shape in particular muscle pairs (Fig. 1) we accepted MAMAD, MAMADdes and MAMADasc as the measures of homologous muscle symmetry.

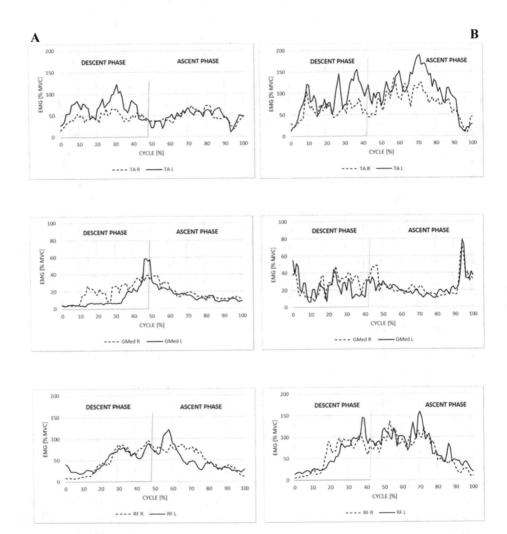

Fig. 1. EMG profiles normalized according to amplitude and time [% MVC] for three pairs of homologous lower limb muscles (TA – *tibialis anterior*, G_{med} – *gastrocnemius mediale*, RF – *rectus femoris*) of a representative subject. Squatting attempts (in the descent and ascent phases) were performed with the following loads: (A) 70% 1RM and (B) 100% 1RM [8].

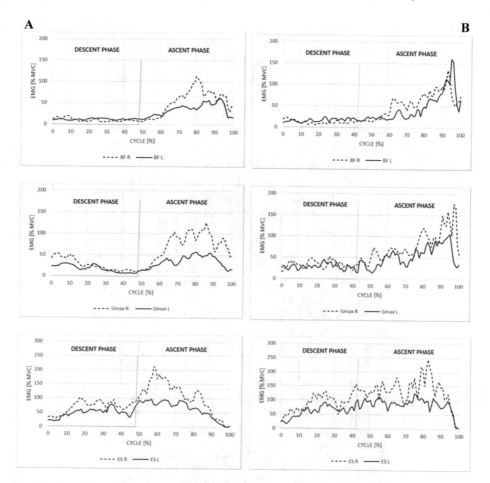

Fig. 1. (Cont.) EMG profiles normalized according to amplitude and time [% MVC] for three successive pairs of homologous lower limb muscles (BF – *biceps femoris*, G_{max} – *gluteus maximus*, ES – *erector spinae*) of a representative subject.

Table 2. Mean of **averages** of **modules** of **amplitude differences** – MAMAD [% MVC] for the pairs of homologous lower limb muscles. Squatting attempts were performed with 70, 80, 90, and 100% 1RM (one repetition maximum) loads. Explanation of symbols in the text.

Load muscle	70%		80%		90%		100%	
	Mean	± SD	Mean	± SD	Mean	± SD	Mean	± SD
TA	15.75	11.64	14.68	13.75	18.45	16.08	20.76	18.49
G_{med}	8.79	6.46	9.89	7.77	8.80	6.92	8.74	6.73
RF	14.22	11.27	13.08	11.33	12.94	9.91	13.32	10.97
BF	8.74	7.66	11.62	9.30	10.73	8.23	10.01	8.32
G_{max}	28.28	27.72	30.10	25.97	36.15	31.28	32.47	31.45
ES	15.64	12.03	18.84	12.52	27.18	14.40	21.52	16.49

Table 3. Comparative analysis of load relationships in the squat. * Significant at $p < 0.05$.

A pair of variables	Wilcoxon test	P
TA 70% 1RM; TA 80% 1RM	0.845	0.398
TA 70% 1RM; TA 90% 1RM	1.014	0.310
TA 70% 1RM; TA 100% 1RM	1.183	0.237
TA 80% 1RM; TA 90% 1RM	2.028	0.043*
TA 80% 1RM; TA 100% 1RM	2.197	0.028*
TA 90% 1RM; TA 100% 1RM	1.183	0.237
G_{med} 70% 1RM; G_{med} 80% 1RM	0.338	0.735
G_{med} 70% 1RM; G_{med} 90% 1RM	0.000	1.000
G_{med} 70% 1RM; G_{med} 100% 1RM	0.338	0.735
G_{med} 80% 1RM; G_{med} 90% 1RM	0.338	0.735
G_{med} 80% 1RM; G_{med} 100% 1RM	0.676	0.499
G_{med} 90% 1RM; G_{med} 100% 1RM	0.000	1.000
RF 70% 1RM; RF 80% 1RM	0.338	0.735
RF 70% 1RM; RF 90% 1RM	1.014	0.310
RF 70% 1RM; RF 100% 1RM	0.169	8.666
RF 80% 1RM; RF 90% 1RM	0.169	0.866
RF 80% 1RM; RF 100% 1RM	0.507	0.612
RF 90% 1RM; RF 100% 1RM	0.507	0.612
BF 70% 1RM; BF 80% 1RM	1.183	0.237
BF 70% 1RM; BF 90% 1RM	0.845	0.398
BF 70% 1RM; BF 100% 1RM	0.507	0.612
BF 80% 1RM; BF 90% 1RM	0.169	0.866
BF 80% 1RM; BF 100% 1RM	0.338	0.735
BF 90% 1RM; BF 100% 1RM	0.338	0.735
G_{max} 70% 1RM; G_{max} 80% 1RM	1.183	0.238
G_{max} 70% 1RM; G_{max} 90% 1RM	1.352	0.176
G_{max} 70% 1RM; G_{max} 100% 1RM	0.845	0.398
G_{max} 80% 1RM; G_{max} 90% 1RM	1.352	0.176
G_{max} 80% 1RM; G_{max} 100% 1RM	1.014	0.310
G_{max} 90% 1RM; G_{max} 100% 1RM	0.000	1.000
ES 70% 1RM; ES 80% 1RM	1.014	0.310
ES 70% 1RM; ES 90% 1RM	2.366	0.018*
ES 70% 1RM; ES 100% 1RM	2.366	0.018*
ES 80% 1RM; ES 90% 1RM	1.352	0.176
ES 80% 1RM; ES 100% 1RM	1.521	0.128
ES 90% 1RM; ES 100% 1RM	1.352	0.176

The calculated MAMAD values for particular homologous muscle pairs in trials with 70, 80, 90 and 100% 1RM loads are presented in Table 2. Data from Table 2 indicate that only three homologous muscle pairs were characterized by higher asymmetry (higher MAMAD values) during the full squat when the weight of the barbell was increased. These were: TA, G_{max} and ES, although there were exceptions.

However, as shown in Table 3, statistically significant higher values in the higher load trials were found in only four of the 36 cases that were checked.

Table 4 presents the calculated MAMAD values for individual homologous muscle pairs in the descent and ascent phases in tests with 70, 80, 90 and 100% 1RM loads.

Table 4. Mean of **averages** of **modules** of **amplitude differences** [% MVC] for the pairs of homologous muscles in the descent and ascent phases (MAMAD$_{des}$ and MAMAD$_{asc}$, respectively) of the squat movement. Squatting attempts were performed with 70, 80, 90, and 100% 1RM loads. Explanation of symbols in the text.

Load Muscle		70% Descent	Ascent	80% Descent	Ascent	90% Descent	Ascent	100% Descent	Ascent
TA	Mean	16.76	8.70	15.15	14.19	20.03	16.64	18.54	22.63
	± SD	11.30	8.29	11.26	13.36	14.07	14.86	14.26	18.74
G_{med}	Mean	8.15	9.53	10.65	9.06	7.41	9.20	9.03	8.56
	± SD	4.91	7.37	8.26	6.23	5.45	7.33	6.64	6.73
RF	Mean	11.22	17.12	13.57	12.89	12.27	13.59	13.40	15.59
	± SD	7.43	12.67	10.14	12.18	8.11	10.51	10.64	10.55
BF	Mean	5.59	11.77	11.19	11.44	6.28	13.75	6.38	13.30
	± SD	3.18	8.87	6.74	8.40	3.53	9.73	4.24	9.30
G_{max}	Mean	13.69	42.76	16.21	42.80	16.02	48.68	18.29	45.38
	± SD	10.81	29.90	13.16	29.04	13.82	34.69	13.24	35.49
ES	Mean	12.60	18.61	15.92	18.99	19.36	23.94	21.24	21.58
	± SD	8.64	13.42	11.71	13.02	1.45	14.81	14.84	15.92

For three muscles (BF, G_{max}, and ES), the ascent phase of the squat was accompanied by statistically significant larger MAMAD values compared to the descent phase, with some exceptions. This was indicated by the Wilcoxon test shown in Table 5. There was no statistically significant difference in the measure of muscle symmetry between the ascent and descent phases for 70% 1RM load for the BF muscle and 90% and 100% for the ES muscle.

Table 5. Comparative analysis of phase relationships in the squat. * Significant at $p < 0.05$.

A pair of variables	Load size	Wilcoxon test	P
TA descent; TA ascent	70% 1RM	1.690	0.091
TA descent; TA ascent	80% 1RM	0.169	0.866
TA descent; TA ascent	90% 1RM	0.676	0.499
TA descent; TA ascent	100% 1RM	0.845	0.398
G_{med} descent; G_{med} ascent	70% 1RM	0.676	0.499
G_{med} descent; G_{med} ascent	80% 1RM	1.352	0.176
G_{med} descent; G_{med} ascent	90% 1RM	1.183	0.237
G_{med} descent; G_{med} ascent	100% 1RM	0.676	0.499
RF descent; RF ascent	70% 1RM	1.859	0.063
RF descent; RF ascent	80% 1RM	0.169	0.866
RF descent; RF ascent	90% 1RM	0.676	0.499
RF descent; RF ascent	100% 1RM	0.338	0.735
BF descent; BF ascent	70% 1RM	2.366	0.018*
BF descent; BF ascent	80% 1RM	0.169	0.866
BF descent; BF ascent	90% 1RM	2.366	0.018*
BF descent; BF ascent	100% 1RM	2.366	0.018*
G_{max} descent; G_{max} ascent	70% 1RM	2.197	0.028*
G_{max} descent; G_{max} ascent	80% 1RM	2.366	0.018*
G_{max} descent; G_{max} ascent	90% 1RM	2.366	0.018*
G_{max} descent; G_{max} ascent	100% 1RM	2.197	0.028*
ES descent; ES ascent	70% 1RM	2.028	0.043*
ES descent; ES ascent	80% 1RM	2.366	0.018*
ES descent; ES ascent	90% 1RM	1.352	0.176
ES descent; BF ascent	100% 1RM	0.169	0.866

4 Discussion

In the past the EMG signals of some homologous leg and hip muscles have been repeatedly recorded [10, 12, 19] but there was no quantitative comparison of EMG profiles for symmetry. Only Arsenault et al. [1] studied symmetry when there was a high correlation between the shapes of whole waveforms of previously rectified and filtered EMG signals for the homologous muscles *rectus femoris* and *soleus* in adults while walking. Since temporal and distance factors of gait exhibit reasonable symmetry [11], it is tempting to assume EMG symmetry for homologous muscles, although this assumption has not been fully explored, and relatively little research has objectively considered symmetry.

In our study the differences in symmetry of six pairs of homologous leg and hip muscles during a squat with increasing loads of the barbell were demonstrated. A statistically significant increase of symmetry of EMG profiles (MAMAD) with increasing loads was seen only for TA and ES muscles and only in two of the six of analyzed cases (Table 3). In the other muscles this trend was not found. Of particular interest is the statistically significant large MAMAD increase for BF and ES muscles and for the so called prime mover (G_{max}; [15]) in the ascent phase of the squat compared to the descent phase (Table 4). In squat studies on participants with damaged ACL and with no loads [18] differences in the EMG activity of homologous muscles between injured and non-injured sides were also seen.

The kind of ensemble envelopes (EMG profiles) for each pair of homologous muscles from individual subjects was determined by calculating the modulus of the difference in the amplitude values at each of the 100 time-normalized points. The grand ensemble averages [14, 15, 20] were created for each muscle and each activity using the ensemble envelopes for all subjects. Examples of the grand ensemble averages for the BF muscle while squatting with 70 and 100% 1RM loads are shown in Fig. 2. The grand ensemble average (mean ± 1 SD) of the BF muscle of all participants differed from the curves of individual participants (Fig. 1). Õunpuu and Winter [13] concluded that whereas pooled data (the grand ensemble averages) reflected reasonable symmetry based on statistical analysis, such data concealed asymmetry (bilateral differences) in individual participants. Therefore, in our last paper [8] we conducted a case study. In this paper we found that increased squat loads increased the asymmetry of the homologous muscle profiles of the lower limbs. However, a few exceptions were also registered. It is particularly interesting that less fluidity of motion accompanied the large asymmetry of major leg muscles during the ascent phase of the squat at a 100% 1RM load. This occurred because there was an altered movement pattern during the ascent phase caused by the knee joints and the hip girdle rotations, once in one direction and then in the opposite. However, it is not possible to carry out such detailed analysis for the grand ensemble averages.

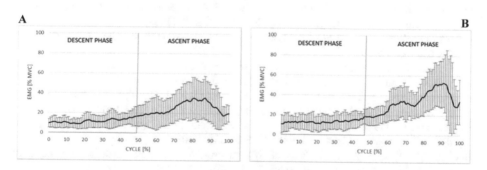

Fig. 2. Grand ensemble averages (± SD) of the BF muscle for all 7 participants while squatting with the following loads: (A) 70% 1RM and (B) 100% 1RM.

References

1. Arsenault, A.B., Winter, D.A., Marteniuk, R.G.: Bilateralism of EMG profiles in human locomotion. Am. J. Phys. Med. **65**, 1–16 (1986)
2. Beachle, T.R., Earle, R.W., Wathen, D.: Resistance training. In: Baechle, T.R., Earle, R.W. (eds.) Essentials of Strength Training and Conditioning. Champaign, IL, pp. 381–412. Human Kinetics (2008)
3. Contreras, B., Vigotsky, A.D., Schoenfeld, B.J., Beardsley, C., Cronin, J.: A comparison of gluteus maximus, biceps femoris, and vastus lateralis EMG amplitude in the parallel, full, and front squat variations in resistance trained females. J. Appl. Biomech. **32**(1), 16–22 (2016)
4. Delavier, F.: Strength Training Anatomy. Human Kinetics, Champaign (2001)
5. Gullett, J.C., Tillman, M.D., Gutierrez, G.M., Chow, J.W.: A biomechanical comparison of back and front squats in healthy trained individuals. J. Strength Cond. Res. **23**(1), 284–292 (2009)
6. Hermens, J., Freriks, B., Merletti, R., Stegman, D., Blok, J., Rau, G., Disselhorst-Klug, C., Hägg, G., et al.: SENIAM 8: European Recommendations for Surface Electromyography. Roessingh Research and Development B.v., The Netherlands (1999)
7. Ingersoll, C.D., Grindstaff, T.L., Pietrosimone, B.G., Hart, J.M.: Neuromuscular consequences of anterior cruciate ligament injury. Clin. Sports Med. **27**(3), 383–404 (2008)
8. Kmiecik, K., Król, H., Sobota, G.: Are lower limb electromyogram profiles symmetrical during a barbell squat? J. Kinesiol. Exerc. Sci. **78**(27), 65–74 (2017). (A case study). Antropomotoryka
9. Konrad, P.: ABC of EMG. A Practical Introduction to Kinesiological Electromyography. Version 1.0, Noraxon Inc., Scottsdale, Arizona, USA (2006)
10. Lyons, K., Perry, J., Gronley, J.K., Barnes, L., Antonelli, D.: Timing and relative intensity of hip extensor and abductor muscle action during level and stair ambulation: an EMG study. Phys. Ther. **10**, 1597–1605 (1983)
11. Murray, M.P., Mollinger Gardner, G.M., Sepic, S.B.: Kinematic and EMG patterns during slow, free, and fast walking. J. Orthop. Res. **2**, 272–280 (1984)
12. Nashner, L.M.: Balance adjustments of humans perturbed while walking. J. Neurophysiol. **44**, 650–664 (1980)
13. Õunpuu, S., Winter, D.A.: Bilateral electromyographical analysis of the lower limbs during walking in normal adults. Electroencephalogr. Clin. Neurophysiol. **72**, 429–438 (1989)
14. Pierotti, S.E., Brand, R.A., Gabel, R.H., Pedersen, D.R., Clarke, W.R.: Are leg electromyogram profiles symmetrical? J. Orthop. Res. **9**, 720–729 (1991)
15. Robertson, D.G.E., Wilson, J.M.J., St. Pierre, T.A.: Lower extremity muscle functions during full squats. J. Appl. Biomech. **24**, 333–339 (2008)
16. Rudolph, K.S., Snyder, L.: Effect of dynamic stability on a step task in ACL deficient individuals. J. Electromyogr. Kinesiol. **14**(5), 565–575 (2004)
17. Trulsson, A., Garwicz, M., Ageberg, E.: Postural orientation in subjects with anterior cruciate ligament injury: development and first evaluation of a new observational test battery. Knee Surg. Sports Traumatol. Arthroscopy **18**(6), 814–823 (2010)
18. Trulsson, A., Miller, M., Hansson, G.-A., Gummesson, C., Garwicz, M.: Altered movement patterns and muscular activity during single and double leg squats in individuals with anterior cruciate ligament injury. BMC Musculoskelet. Disord. **6**(1), 472–482 (2015)
19. Woltering, H., Güth, V., Abbink, F.: Electromyographic investigations of gait in cerebral palsied children. Electromyogr. Clin. Neurophysiol. **19**, 519–533 (1979)

20. Yang, J.F., Winter, D.: Electromyographic amplitude normalization method: improving their sensitivity as diagnostic tools in gait analysis. Arch. Phys. Med. Rehabil. **65**, 517–521 (1984)
21. Yang, J.F., Winter, D.: Surface EMG profiles during different walking cadences in humans. Electroencephalogr. Clin. Neurophysiol. **60**(6), 485–491 (1985)
22. Yavuz, H.U., Erdag, D.: Kinematic and electromyographic activity changes during back squat with submaximal and maximal loading. Appl. Bionics Biomech. **2017**, 8 (2017)

Does Vibration Affect Upper Limb During Nordic Walking?

Wojciech Wolański[1(✉)], Michał Burkacki[1], Sławomir Suchoń[1], Julia Gruszka[1], Marek Gzik[1], Krzysztof Gieremek[2], and Joanna Gorwa[3]

[1] Faculty of Biomedical Engineering, Department of Bimoechatronics, Silesian University of Technlogy, ul. Roosevelta 40, 41-800 Zabrze, Poland
wwolanski@polsl.pl
[2] Department of Physical Therapy and Therapeutic Massage, The Jerzy Kuczka Academy of Physical Education, ul. Mikołowska 72a, 40-065 Katowice, Poland
[3] Faculty of Physical Education, Sport and Rehabilitation, Department of Biomechanics, Poznan University of Physical Education, ul. Królowej Jadwigi 27/39, 61-871 Poznań, Poland
https://www.polsl.pl/Wydzialy/RIB/RIB3en/Strony/welcome.aspx

Abstract. Nordic Walking has become a very popular physical activity. The technique is relatively simple and Nordic Walking poles are cheap and accessible. However, it should be noted that Nordic Walking generates additional loads for upper limb joints in the form of mechanical vibrations, which may not be without effect on exercisers' health. The aim of this paper is to address the following questions: Does Nordic Walking cause harmful mechanical vibrations? Does the use of poles with amortization reduce the level of vibration? May the mechanical vibration during this exercise be a contraindication to practise Nordic Walking? To answer these questions, the authors used a vibration measurement device, which allowed the researchers to measure vibration according to norm EN ISO 5349 during gait with Nordic Walking poles. Data collected for different types of poles were compared to the regulation of the Polish Minister of Labour and Social Policy 2002.

Keywords: Ergonomics · Nordic walking · Vibrations · Joint load

1 Introduction

Nordic Walking has become a very popular physical activity. The walking technique is relatively simple and Nordic Walking poles are cheap and accessible. Nordic Walking is commonly recommended to everyone as a training method with low requirements [1]. Nordic Walking has a good effect, for example, in a hypertension reduction [2], improvement of venous blood flow in lower extremities [3] or obesity management [4]. Another alleged advantage of Nordic Walking is the reduction of loads acting on the lumbar spine, hip and knee joint [5–7]. However some of the above-mentioned advantages are disputed. Recent studies

© Springer Nature Switzerland AG 2019
K. Arkusz et al. (Eds.): BIOMECHANICS 2018, AISC 831, pp. 276–284, 2019.
https://doi.org/10.1007/978-3-319-97286-2_25

[8, 9] shows that Nordic Walking training results in the increase of ground reaction forces in the phase involving step forward on the heel. It was shown in [10] that Nordic Walking relieves lower limbs at the cost of upper limb involvement. Study presented in [10] also shows that Nordic Walking is less energy efficient in comparison to walking. Taking into consideration these facts, a question must be asked if there is possibility that intensive Nordic Walking training could have negative effects on upper limbs? There is a lack of information on vibration loads acting on upper limbs, especially wrists. Therefore, the aim of this paper is to answer the following questions: Do Nordic Walking poles cause harmful mechanical vibrations? Does the use of poles with amortization reduce the level of vibration? May the mechanical vibration be a contraindication to practise Nordic Walking?

2   Materials and Methods

The main objects of the research presented in this article are Nordic Walking poles. Four types of sticks (as described in Table 1) were tested in three surface conditions: grass, asphalt and floor tiles. Three persons familiar with Nordic Walking technique took part in the tests. There were two women and one man, aged 25.

During the tests, upper limb kinematics and wrist vibrations were recorded simultaneously. One test contained several measurement cycles, one for each footstep. Further analysis was conducted in the sections of 15 footsteps (approximately, a distance of 11 m). Overall 540 samples were used for the determination of vibration exposure. An example of a data plot and statistical analysis of data are presented in Figs. 5 and 6 as well as in Tables 2 and 3.

2.1   Experimental Conditions

Each subject performed Nordic Walking in the conditions of rigid surface (grass, asphalt and floor tiles). A cadence (rhythm) of walk was imposed for the subject by a metronome set at a frequency of 1 Hz. It enabled the obtainment of unified data. The subject's kinematics was registered using the MVN Xsense 3D motion capture suit in upper-body configuration (Fig. 1). It is popular system with high measurement accuracy which is often used in sport activity researches [11] or for determining of correctness of upper limbs motions [12]. The MVN suit makes it possible to determine kinematic values like joint angle or limbs trajectory during walk. Kinematic values are measured by 10 tracker sensors set at a sampling rate of 120 Hz. Unfortunately, the motion capture suit is unsuitable for measuring the wrist acceleration with required parameters. For this purpose, a special wrist vibration measurement system was developed. The device incorporates two ADXL335 3 directional accelerometers and Atmega processor board for data acquisition (Fig. 2). Both accelerometers were attached to the subject wrist and vibration parameters were collected during Nordic Walking. The MVN suit as well as accelerometer sensors are lightweight and do not affect experimental conditions.

Fig. 1. MVN Xsense 3D motion capture suit.

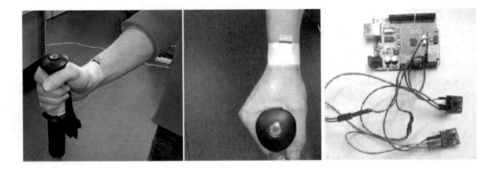

Fig. 2. Accelerometer placement and wrist vibration measurement system.

Table 1. Characteristic of Nordic walking poles used in experimental testing.

	Pole no. 1	Pole no. 2	Pole no. 3	Pole no. 4
Shaft material	Aluminium alloy	Aluminium alloy 5083	Aluminium alloy 7001	Aluminium alloy 7075
Shaft construction	3 segments	1 piece	3 segments	3 segments
Height range	65–135 cm	115 cm	61–132 cm	65–135 cm
Antishock system	Yes	No	Yes	Yes
Handgrip	Plastic	Plastic	Plastic and cork	Plastic and cork
Straps type	Standard	Half-glove	Standard	Standard

2.2   Methods

The authors conducted tests according to the guidelines issued by the Central Institute for Labour Protection and The Polish norm PN-EN ISO 5349. Kinematic and vibration analyses for all subjects were conducted with a view to the following assumptions:

- gait rate - 120 steps per minute,
- single cycle duration - 1 s,
- average step length - 0.75 m,
- distance travelled in 1 s - 1.5 m,
- average velocity: 5.4 km/h.

Data processing was conducted with a prepared script developed in Matlab software. At first, raw data were filtered with a high-pass filter (cut-off frequency $fc = 6.31$ Hz) and a low-pass filter ($fc = 189.5$ Hz) for noise reduction. Then, the signals were filtered with a frequency-weighted filter according to recommendation (ISO 5349). The filters characteristics are presented in Figs. 3 and 4. In the next step, an effective frequency-weighted acceleration was calculated. The effective frequency-weighted acceleration is a basic value used for the evaluation of body exposure to vibrations. A daily exposition value is calculated by means of the identification of vibrations and exposure time. Because Nordic Walking training does not last 8h (work health and safety regulations) the daily exposure was calculated for 1h. Obtained values were confronted with limit values provided by the Central Institute for Labour Protection. A limit value of

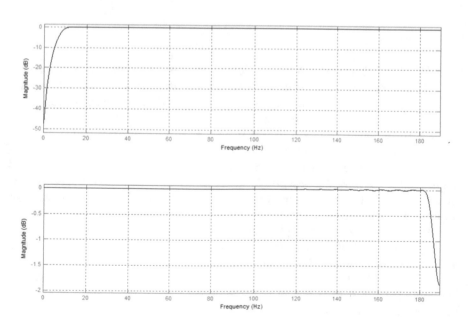

Fig. 3. High-pass filter (cut-off frequency $fc = 6.31$ Hz) and low-pass filter ($fc = 189.5$ Hz) for noise reduction.

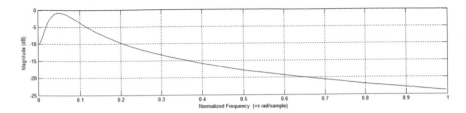

Fig. 4. Frequency-weighted filter.

daily exposure to the mechanical vibrations of upper-arm is $A(h) = 2.8 \text{ m/s}^2$ calculated according to the below-presented formula [13, 14]:

$$A(h) = \sqrt{\frac{1}{T} \sum_{i}^{n} a_{hv_i}^2 t_i} \tag{1}$$

where:

h - exposure time (8 h for labour norms, 1h for NW training);
$a_h v_i$ - total vibration value for i-th operation;
n - number of individual upper limb exposures to vibration;
T - total exposure time during the day;
t_i - duration of i-th exposure.

3   Results

Vibration analysis allows the calculation of weighted acceleration. For example, the acceleration during walking for each person with pole 3, used for exposure calculation, was presented in Fig. 5 and Table 2. Obtained acceleration values for all poles (Table 3) were compared to limit values (Table 4). This made it possible to determine the impact of a pole type on the wrist joint. Figures 5 and 6 illustrate how a frequency filter works: low-frequency signal components are decreased and high frequency signal components are amplified. In Table 2, single samples of duration and acceleration (after weighted filtering process) are presented. Table 3 shows the statistical analysis of samples' average duration and amplitude in the performed test.

Final results, the upper limb exposure to vibrations during Nordic Walking training calculated with formula 1, are shown in Table 4 as percentage of the limit value. Exposure to vibrations exceeded the limit value only in 6 out of 36 cases. Most of them were observed for pole no. 1, mainly on grass and asphalt. The limit value was exceeded once for pole no. 4, whereas it was not exceeded for poles no. 2 and no. 3.

Table 2. Example data: samples (each step - i-th operation) of test with pole no. 3 on asphalt for all persons.

Sample (step)	Person no. 1 t [ms]	Person no. 1 a [m/s2]	Person no. 2 t [ms]	Person no. 2 a [m/s2]	Person no. 3 t [ms]	Person no. 3 a [m/s2]
1	43	7,17	49	6,8	57	4,73
2	45	4,39	47	9,07	33	6,21
3	42	6,24	59	8,26	39	7,75
4	45	6,05	41	8,91	65	6,82
5	50	5,12	34	14,67	41	6,48
6	48	5,32	46	10	40	6,36
7	46	5,27	45	8,27	40	5,41
8	48	5,48	41	7,41	42	5,36
9	51	5,29	50	7,17	49	4,46
10	55	491	42	9,21	48	4,64
11	56	5,91	32	8,93	48	5,19
12	75	4,57	37	7,82	46	5,81
13	54	4,51	42	7,29	46	5,74
14	55	5,35	43	5,48	37	4,46
15	59	4,55	53	7,31	47	5,96

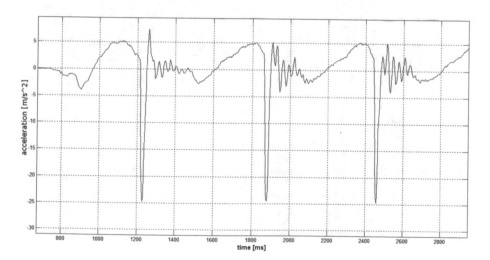

Fig. 5. Pole no. 3, asphalt, person no1; acceleration during walking before the application of frequency-weighted filter.

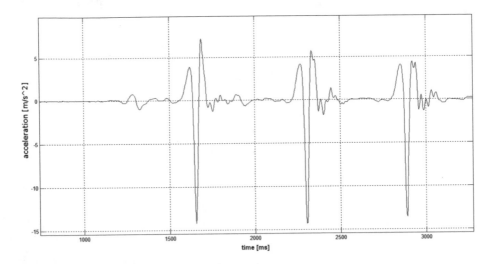

Fig. 6. Pole no. 3, asphalt, person no1; acceleration during walking after the application of frequency-weighted filter.

Table 3. Average time of sample $t_a v$ $[ms]$ and average acceleration $a_a v$ $[m/s^2]$ in tests

		Pole no. 1	Pole no. 2	Pole no. 3	Pole no. 4
Person no. 1	grass	$t_a v = 53.27\ (8.19)$ $a_a v = 9.33\ (1.21)$	$t_a v = 57.33\ (12.96)$ $a_a v = 5.99\ (1.13)$	$t_a v = 56.60\ (13.05)$ $a_a v = 5.88\ (1.22)$	$t_a v = 51.87\ (2.61)$ $a_a v = 7.64\ (1.40)$
	asphalt	$t_a v = 45.13\ (4.44)$ $a_a v = 9.96\ (1.64)$	$t_a v = 49.33\ (6.34)$ $a_a v = 5.14\ (0.74)$	$t_a v = 51.47\ (8.30)$ $a_a v = 37.75\ (125.39)$	$t_a v = 50.33\ (7.40)$ $a_a v = 6.83\ (1.52)$
	tiles	$t_a v = 55.73\ (8.33)$ $a_a v = 4.42\ (1.10)$	$t_a v = 49.93\ (3.56)$ $a_a v = 2.79\ (0.50)$	$t_a v = 47.20\ (9.69)$ $a_a v = 4.35\ (0.85)$	$t_a v = 39.80\ (12.21)$ $a_a v = 5.22\ (1.16)$
Person no. 2	grass	$t_a v = 49.53\ (4.90)$ $a_a v = 9.30\ (1.82)$	$t_a v = 39.67\ (8.60)$ $a_a v = 9.68\ (1.75)$	$t_a v = 47.53\ (10.63)$ $a_a v = 8.42\ (2.13)$	$t_a v = 46.87\ (8.25)$ $a_a v = 8.89\ (1.67)$
	asphalt	$t_a v = 53.2\ (10.86)$ $a_a v = 10.49\ (3.45)$	$t_a v = 49.53\ (9.74)$ $a_a v = 7.17\ (1.04)$	$t_a v = 44.07\ (7.08)$ $a_a v = 8.44\ (2.07)$	$t_a v = 50.20\ (7.24)$ $a_a v = 9.46\ (2.00)$
	tiles	$t_a v = 38.27\ (11.20)$ $a_a v = 5.70\ (0.50)$	$t_a v = 50.80\ (2.74)$ $a_a v = 5.61\ (0.53)$	$t_a v = 49.90\ (8.43)$ $a_a v = 6.64\ (0.60)$	$t_a v = 42.17\ (7.58)$ $a_a v = 9.06\ (3.21)$
Person no. 3	grass	$t_a v = 53.07\ (8.54)$ $a_a v = 8.98\ (1.19)$	$t_a v = 45.07\ (6.16)$ $a_a v = 5.21\ (0.86)$	$t_a v = 40.00\ (4.38)$ $a_a v = 7.55\ (1.48)$	$t_a v = 40.00\ (7.04)$ $a_a v = 8.21\ (1.91)$
	asphalt	$t_a v = 43.87\ (5.03)$ $a_a v = 9.05\ (1.29)$	$t_a v = 45.93\ (5.74)$ $a_a v = 5.19\ (0.83)$	$t_a v = 45.20\ (8.03)$ $a_a v = 5.69\ (0.94)$	$t_a v = 45.07\ (5.69)$ $a_a v = 6.44\ (1.32)$
	tiles	$t_a v = 50.43\ (8.30)$ $a_a v = 5.36\ (0.58)$	$t_a v = 48.80\ (5.67)$ $a_a v = 2.29\ (0.68)$	$t_a v = 44.73\ (7.89)$ $a_a v = 4.76\ (0.79)$	$t_a v = 43.07\ (4.95)$ $a_a v = 4.72\ (0.79)$

Table 4. Exposure to vibrations as percentage of limit value

		Pole no. 1	Pole no. 2	Pole no. 3	Pole no. 4
Person no. 1	grass	109	71	70	89
	asphalt	108	58	61	79
	tiles	53	32	49	53
Person no. 2	grass	106	96	91	97
	asphalt	127	80	90	108
	tiles	56	64	75	95
Person no. 3	grass	104	56	72	82
	asphalt	96	56	62	69
	tiles	61	26	51	50

4   Discussion

The conducted studies emphasized new aspects of Nordic Walking and derived important conclusions. The authors observed that a vibration exposition dose during Nordic Walking in some tests exceeded the limit value of $A(h) = 2.8\ \mathrm{m/s^2}$ according to regulations and may have a negative outcome for a training person. The maximum exposure value was 127% of the safe limit value. Research has shown that five out of six off-limit values were obtained with pole no. 1, despite an Antishock system. On the other hand, pole without amortization does not exceed limit value. This means that the choice of a correct pole type is an essential factor in Nordic Walking. This is also suggested by further studies which are focused on the damping of poles for prevention of vibration transfer. Also, a surface type is an important factor. The smallest exposure values were obtained on the tiles ground; as observed, person no. 2 had higher values than the rest. Obtained results show that 1 h training of Nordic Walking in particular conditions may have a harmful effect on upper limbs. However, the exposure can be reduced by using good poles and by improving a walking technique. The tiles ground has shown the best properties for training despite its rigidness. This may be the effect of weight filtering which reduces high frequencies. Summing up, the vibration of poles transferred to the upper limbs during training depends on many aspects: ground type, pole model, knowledge of techniques and physical differences between individuals. In some combinations of these conditions, Nordic Walking training may produce a negative outcome to health. However, these observations corresponding to the studies of [15] may be seen as an advantage. Some vibrations with particular parameters may have a positive effect and become useful in specific, non-typical training methods.

References

1. Kantaneva, M.: Original Nordic Walking e-book (2005)
2. Latosik, E., Zubrzycki, I.Z., Ossowski, Z., Bojke, O., Clarke, A., Wiacek, M., Trabka, B.: Physiological responses associated with nordic-walking training in systolic hypertensive postmenopausal women. J. Hum. Kinet. **43**, 185–190 (2014). https://doi.org/10.2478/hukin-2014-0104
3. Jasiński, R., Socha, M., Sitko, L., Kubicka, K., Woźniewski, M., Sobiech, K.A.: Effect of nordic walking and water aerobics training on body composition and the blood flow in lower extremities in elderly women. J. Hum. Kinet. **45**, 113–122 (2015). https://doi.org/10.1515/hukin-2015-0012
4. Figard-Fabre, H., Fabre, N., Leonardi, A., Schena, F.: Efficacy of nordic walking in obesity management. Int. J. Sports Med. **32**(6), 407–414 (2011)
5. Koizumi, T., Tsujiuchi, N., Takeda, M., Murodate, Y.: Physical motion analysis of Nordic walking. In: The Engineering of Sport 7, vol. 1, pp. 336–385 (2008)
6. Koizumi, T., Tsujiuchi, N., Takeda, M., Fujikura, R., Kojima, T.: Load dynamics of joints in Nordic walking. In: 5th Asia-Pacific Congress on Sports Technology (APCST) (2011). https://doi.org/10.1016/j.proeng.2011.11.2750
7. Wendlova, J.: Nordic walking - is it suitable for patients with fractured vertebra? Bratislavske lekarske listy **109**(4), 171–176 (2008)
8. Dziuba, A.K., Żurek, G., Garrard, I., Wierzbicka-Damska, I.: Biomechanical parameters in lower limbs during natural walking and Nordic walking at different speeds. Acta Bioeng. Biomech. **17**(1), 95–101 (2015). https://doi.org/10.5277/ABB-00077-2014-01
9. Cieśla, W., Gieremek, K., Drabik, J., Gorny, M.: Pedobarographic analysis of foot load during gait with Nordic walking poles. Med. Sportowa **31**(3), 129–135 (2015). GICID: 01.3001.0008.4459
10. Pellegrinia, B., Peyré-Tartarugac, L.A., Zoppirollia, Ch., Bortolana, L., Savoldellia, A., Minettid, A.E., Schenaa, F.: Mechanical energy patterns in Nordic walking: comparisons with conventional walking. Gait Posture **51**, 234–238 (2017)
11. Michnik, R., Wodarski, P., Bieniek, A., Jurkojć, J., Mosler, D., Kalina, R.M.: Effectiveness of avoiding collision with an object in motion - virtual reality technology in diagnostic and training from perspective of prophylactic of body injuries. Arch. Budo **13**, 203–210 (2017)
12. Jurkojć, J., Wodarski, P., Michnik, R., Nowakowska, K., Bieniek, A., Gzik, M.: The upper limb motion deviation index: a new comprehensive index of upper limb motion pathology. Acta Bioeng. Biomech. **19**(2), 175–185 (2017)
13. Regulation of Polish Minister of Labor and Social Policy 2002 (Rozporzadzenie z dnia 29 listopada 2002 r. w sprawie najwyższych dopuszczalnych steżeń i nateżeń czynników szkodliwych dla zdrowia w środowisku pracy. DzU nr 217, poz. 1833; zm. DzU 2005, nr 212, poz. 1769)
14. Kowalski, P.: Measurement and evaluation of vibration at the workplace according to new regulations. Bezpieczeństwo Pracy, **9**, 24–26 (2006)
15. Chmielewska, D., Piecha, M., Błaszczak, E., Król, P., Smykla, A., Juras, G.: The effect of a single session of whole-body vibration training in recreationally active men on the excitability of the central and peripheral nervous system. J. Hum. Kinet. **41**, 89–98 (2014). https://doi.org/10.2478/hukin-2014-0036

Loaded Treadmill Training Improves the Spatio-Temporal Parameters in Children with Spastic Diplegia

Mariam A. Ameer and Walaa S. Mohammad[✉]

Department of Biomechanics, Faculty of Physical Therapy,
Cairo University, Giza, Egypt
walaa.sayed@pt.cu.edu.eg

Abstract. Background: Treadmill training is a commonly used and promising technique for improving gait function in children with spastic diplegia. However, the use of loads during treadmill gait training is limited. The purpose of the present study was to determine whether using loads with treadmill training improves the spatio-temporal parameters of gait in children with spastic diplegia more effectively than conventional exercises alone. Methods: Twenty children with spastic diplegia were randomly allocated to a control group or an experimental group. Both groups received conventional therapeutic exercises for a period of eight weeks. Moreover, the experimental group underwent loads on their ankles during treadmill gait training. Spatio-temporal parameters of children were assessed at baseline (pre-training) and at their 16th training session (post-training). Results: A two-way mixed-design ANOVA showed no significant between-group differences in demographic and the spatio-temporal parameters at baseline. Based on measurements taken at the 16th training session, the experimental group achieved significantly ($p < 0.05$) higher average scores than the control group with regard to step length, stride length, cadence, walking velocity, stride time, and double support time. In addition, the results revealed significant within-group differences ($p < 0.05$) in the step and stride lengths of both groups, whereas cadence, walking velocity, stride time, and double support time also improved in the experimental group. Conclusions: Eight weeks of loaded treadmill gait training improved the gait kinematics of children with spastic diplegia, particularly cadence, walking velocity, stride time, and double support time.

Keywords: Gait training · Load · Spastic diplegia · Spatio-temporal

1 Introduction

Spastic diplegia is a form of cerebral palsy (CP), and is characterized by increased muscle tone, abnormal equilibrium reactions, muscle weakness, and relative imbalance of muscle forces across the joints in the lower limbs [1]. Such abnormalities result in crouched gait, or diplegic gait, which is a common problem characterized by predictable kinematic and kinetic alterations [2]. These changes may result in a slow walking velocity, a shorter step length, a shorter stride length, decreased dorsiflexion

© Springer Nature Switzerland AG 2019
K. Arkusz et al. (Eds.): BIOMECHANICS 2018, AISC 831, pp. 285–293, 2019.
https://doi.org/10.1007/978-3-319-97286-2_26

during the swing phase, a shorter duration of single-limb support, and a longer duration of double support [1–4].

Many researchers have demonstrated the strong positive relationship between lower-limb (LL) muscle strength and ambulatory function in children with spastic diplegia [5, 6]. Encouraging results indicate that increases in LL muscle strength could improve gait and function in persons with spastic diplegia [7–10]. Accordingly, improving the gait efficiency of children with spastic diplegia involves the use of intervention techniques that increase the force experienced during walking. This is achieved by targeting specific muscle groups at ideal times during the gait cycle [11]. In addition, functional strengthening exercises, as well as treadmill and balance training, have shown positive effects on gait in these children [12, 13].

Moreover, therapy for children vs. specific kinds of therapy tested in adults. After the addition and subsequent removal of LL loads (weights) during the swing phase, both healthy adults and adults with neurological pathologies make compensatory adjustments in their gaits immediately after the removal of the imposed resistance [14–17]. Locomotor adjustments include increased walking velocity [17], flexor muscle activity, flexor muscle torque, and height of steps, as well as increases in the angle of hip and knee flexion during the swing phase [15, 16]. The published literature lacks clinical studies investigating the locomotor behavior of children with spastic diplegia in response to load training, via the ankles, during treadmill walking

In light of recognizing these improvements caused by the imposition of LL resistance in gait, we hypothesized that the resulting locomotor adjustments may improve the spatio-temporal variables of the paretic LL during the swing phase of the gait of children with spastic diplegia that is unique in this population. Therefore, the present study was designed to explore the changes in spatio-temporal variables in children with spastic diplegia after gait training using a treadmill and loads around their ankles.

2   Methods

2.1   Sample Characteristics

Twenty participants (9 boys and 11 girls, aged 5–8 years) with spastic diplegia were recruited to participate as volunteers in this study. The children were able to walk without aid devices and were therefore classified as Gross Motor Function Classification system level I or II [18]. The following inclusion-exclusion criteria were used to select the participants of the study: Initially, diagnosis of spastic diplegia was an inclusion criterion. Children were able to walk without the use of assistive devices, had not received orthopedic surgery at least one year prior to the study, and had not received botulinum toxin type-A injections at least six months prior to the study.

Children and their parents received detailed information about the study before giving his/her signed consent. Prior to participation, written informed consent was obtained from all child participants as well as their parents, in accordance with the ethical standards of the Declaration. The local ethics committee approved the study.

2.2 Experimental Design

The study was designed as a randomized control trial. Group randomization was performed by a statistician who was blinded to the study treatment and procedures. To hide their identities, each subject was identified by a specific identification number, and an SPSS program (version 23) was used to randomly divide the participating children into two groups.

The control group was exposed to conventional physical therapy treatment (CPTT) only, while the experimental group received weighted treadmill gait training in addition to CPTT. Both groups completed 60 min of the CPTT sessions 2 times per week for 8 successive weeks. Each session started with warm-up stretches of the major muscle groups. CPTT included the following exercises: stretching exercises for tight muscles (hip adductors, knee flexors, ankle planter flexors, shoulder adductors, and shoulder internal rotators), strengthening exercises for weak muscles in both LL (hip abductors, knee extensors, and ankle planter flexors), and active free exercises for upper limb muscles. In addition, balance and treadmill training and different daily life activities (i.e., sit-to-stand and sit-up exercises) were implemented.

Participants in the experimental group were allowed a rest period of 30 min before continuing with weighted treadmill gait training. The training protocol on the treadmill was performed for 30 min, 2 times per week, for 8 consecutive weeks with loads constituting 60% of each patient's LL weight attached to each participant's ankles bilaterally with shin guards. According to Simão, Galvão [11], adding loads of 60% of each patient's LL weight improved that patient's spatio-temporal gait parameters. Finally, the speed, which was slow at first, gradually increased up to 60% of the maximal tolerance of each participant, as recorded in the baseline evaluation session.

The following spatio-temporal parameters were investigated in both groups: walking velocity, stride length, step length, cadence, double support time (DLS), and stride time. Computerized 3D motion capture and analysis system was used with 16 retro-reflective spherical markers were attached to specific anatomical landmarks on the lower limbs. All pre- and post-testing trials were conducted by the same examiner to minimize potential inter-rater test errors.

2.3 Procedure

First, the past medical histories and anthropometric characteristics (such as weight, height) of the child participants were obtained. Meanwhile, the spasticity degrees of the LL lower muscles were assessed. Based on the calculation proposed by Jensen [19], the LL mass was determined, and the appropriate loads for each child in the experimental group (corresponding to 60% of his/her LL mass) were calculated and utilized during treadmill training.

Prior to each data collection session, the walking speed for each child was estimated using three familiarization trials. The average walking speed of these trials was considered the baseline speed for the following treatment sessions. Before beginning any intervention, all spatio-temporal variables of interest were recorded and used as baseline measurements for both groups. The markers were placed on the lower extremities of all children by only one examiner. The markers were affixed to the bilateral anterior superior iliac spine (ASIS), the bilateral posterior superior iliac spine (PSIS), the lower one-third of the left thigh, the upper one-third of right thigh (to differentiate the right and left lower limbs), the lateral surface of the knee joint, the upper one-third of the lateral surface of the left tibia, the lower one-third of the lateral surface of the right tibia (to discriminate between the right and left lower limbs), the bilateral lateral malleolus, the bilateral metatarsal head of the second toes, and the bilateral heels. Elastic double adhesive and regular tapes were used to minimize marker movement during landing.

The participants were asked to walk at their normal pace along a 10-m walkway. For each child, six successful walking trials were recorded. Spatio-temporal data were collected on two different occasions, pre- and post-training, for both groups. Finally, only the experimental group underwent weighted gait training using a treadmill with a variety of walking velocity controls (American motion fitness MF 8625).

2.4 Data Collection

Spatio-temporal parameters were measured using a computerized 3D motion capture and analysis system (VICON MX-T10 Motion Analysis System, Oxford Metrics Inc., Oxford, UK; Bonita cameras, 250 frames/s) with ten cameras (to detect segment angular position data at 120 Hz). Sixteen retro-reflective spherical markers were attached to specific anatomical landmarks on the lower limbs, according to the plugin gait model of VICON Nexus software 2.2.1, to capture the spatio-temporal data. The data were then saved as C3D files and exported to VICON Polygon software 3.5 via the eclipse data management interface for analysis.

2.5 Statistical Analysis

Data were analyzed using the Statistical Package for Social Sciences (version 23.0 for Windows; SPSS Inc., Chicago, IL). A two-way mixed design analysis of variance (ANOVA) was used to investigate within-group differences (pre- vs. post-training) and between-group differences (experimental vs. control group) in the following spatio-temporal variables: walking velocity, stride length, step length, cadence, double support time (DLS), and stride time. Statistical significance was set at $p < 0.05$.

3 Results

Twenty-eight children with spastic diplegia were identified as potential participants in the present study. A total of 23 children with spastic diplegia fulfilled the inclusion criteria for participation; however, the parents of 3 children refused to permit participation, resulting in a total sample size of 20 children (Fig. 1). There were no significant differences between the control and experimental groups with regard to their ages, heights, weights, and sex. The participants' descriptive statistics and demographics are presented in Table 1.

To test whether conventional therapy and treadmill training with ankle loading improves gait parameters in children with spastic diplegia more effectively than conventional therapy alone, we assessed a variety of kinetic parameters in patients before

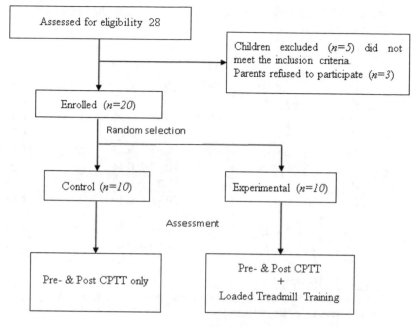

Fig. 1. Study flow chart

Table 1. Demographic data for control and experimental spastic diplegic groups compared by ANOVA.

Groups	Control $n = 10$	Experimental $n = 10$	p value
Age (years)	6.33 ± 0.86	6.49 ± 1.01	0.707
Height (cm)	117.60 ± 12.47	114.80 ± 10.33	0.591
Weight (kg)	35.02 ± 3.83	33.41 ± 4.13	0.378
Sex	5 F, 5 M	6 F, 4 M	0.661

Values are expressed as mean ± SD.
M: Male; F: Female.

and after an 8-week training period. There was a significant main effect of group and a significant interaction between the control and experimental groups. A significant within-group difference was observed between the two time points (pre- and post-training) investigated ($p = 0.000$). There was also a significant between-group differences (interaction effect) between time and the groups tested ($p = 0.000$). *Turkey's Post hoc* test indicated that there were no significant differences (in all parameters) between the two groups at baseline; however, significant differences (in all parameters) exist between both groups at post-training. Spatio-temporal parameters in the control group did not change post-training, except for step and stride lengths. For the experimental group, however, the mean scores of the tested variables (walking velocity, stride length, step length, cadence, DLS, and stride time) improved significantly post-training (Table 2).

Table 2. Mean ± SD of spatiotemporal parameters comparing children's performance in control and experimental groups.

Gait parameters	Control			Experimental		P value[†]
	Mean ± SD		P value	Mean ± SD	P value	
Cadence (step/min)	**Pre**	107.65 ± 6.42	1.000	102.87 ± 7.79	0.000*	0.152
	Post	107.99 ± 6.57		115.00 ± 6.60		0.029*
Step length (m)	**Pre**	0.31 ± 0.06	0.024*	0.34 ± 0.06	0.000*	0.159
	Post	0.37 ± 0.07		0.47 ± 0.07		0.005*
Stride length (m)	**Pre**	0.59 ± 0.12	0.019*	0.68 ± 0.12	0.000*	0.116
	Post	0.71 ± 0.14		0.90 ± 0.14		0.007*
Walking velocity (m/min)	**Pre**	0.75 ± 0.16	1.000	0.76 ± 0.12	0.000*	0.865
	Post	0.78 ± 0.13		0.92 ± 0.14		0.037*
Stride time (s)	**Pre**	1.13 ± 0.07	1.000	1.15 ± 0.09	0.000*	0.532
	Post	1.10 ± 0.07		0.99 ± 0.09		0.017*
DLS (s)	**Pre**	0.45 ± 0.09	1.000	0.44 ± 0.08	0.000*	0.958
	Post	0.47 ± 0.03		0.38 ± 0.09		0.001*

DLS: double support time; SD: standard deviation; Pre: baseline; Post: at 12th session
*$p < 0.05$; † Between-group comparisons.

4 Discussion

Adding loads to the ankles of children with spastic diplegia during treadmill training is unprecedented in this population. We found that the children with spastic diplegia tend to modify their spatio-temporal parameters (walking velocity, cadence, DLS, and stride time), which suggests that they may be able to adapt their locomotor skills in response to additional loads.

Significant between-group improvement in all spatio-temporal parameters investigated were observed post-training. The experimental group demonstrated significant increases in walking velocity and longer step and stride lengths than those of control group. In addition, the cadence of the experimental group increased significantly. This is supposedly because cadence depends on the other spatio-temporal gait parameters, such as velocity [1]. Motor weakness and poor voluntary motor control are characteristics for spastic diplegia which is a type of CP. Therefore, children with CP use a wider base of support than normal children use in order to stabilize their centers of mass [20, 21]. Moreover, previous researcher showed that the step width was correlated with walking velocity, cadence, and stride length. Thus, children with wider step widths tend to have greater gait difficulties [1, 2]. Therefore, the improvement in cadence and walking velocity could explain improvements in children' balance and muscle strength. In the same context, one of the principles in sports, especially for runners during a sprint, is that higher speeds may be achieved at a higher cadence by taking more steps and having more ground support. A higher cadence actually takes less effort, improves forward momentum, decreases injuries, and increases speed [1]. In the same way, the experimental group in our study exhibited improved spatio-temporal gait parameters because of the shorter double support times during their gaits and their longer strides.

Significant within-group improvement in only two parameters investigated were observed in control group. Interestingly, only two parameters (step length and stride length) were significantly higher in children trained with CPTT and this is unexpected given the relationship conventionally seen with other parameters. This is in close agreement with the results of Blundell, Shepherd [12], who reported that the strengthening exercises for LL muscles significantly improved the stride length with non-significant increases in cadence. Similarly, in another study, stride lengths increased significantly after muscle training, with no change in average velocity; these effects were due to better stability around both the hips and the knees in the stance phase [9]. The previous work may explain that the strengthening exercises for LL muscles; which represents CPTT in our study, could improve the spatial parameters (step length and stride length) with no effect on temporal parameter (cadence and velocity).

In contrast, significant within-group improvement in all parameters investigated were observed in experimental group. Loaded treadmill training improves both walking velocity and cadence, which is reflected in children' gait. This may be attributed to the ankle-bearing loads, which have been found to impose resistance throughout the swing phase, causing more activation of LL flexor muscles [15, 16]. In order to explain these adaptive modifications, some authors [15, 22, 23] have proposed that changes in proprioceptive input (the loads) engage feedback strategies that cause neuromotor adaptation. Upon removal of such loads, feed-forward adaptive strategies account for the adaptations that required experience to develop and that continue to manifest during walking.

5   Conclusion

The results of this investigation indicate the importance of loaded gait training in improving the gait parameters of children with spastic diplegia. We found meaningful differences between the experimental and control groups with regard to velocity, stride length, step length, cadence, stride time, and double support time. However, loaded treadmill gait training improves the spatio-temporal parameters of gait; the CPTT impacts only step and stride lengths. In conclusion, the results reported here demonstrate that eight weeks of loaded treadmill gait training improves gait parameters; particularly temporal parameters, in children with spastic diplegia.

Contributions. MA: Conceived the study, data collection, and analysed the data; WSM: modify the data analysis, manuscript preparation, review, submission and will act as guarantor for the paper. All authors approved the final version to be published.

References

1. Pauk, J., Ihnatouski, M., Daunoraviciene, K., Laskhousky, U., Griskevicius, J.: Research of the spatial-temporal gait parameters and pressure characteristic in spastic diplegia children. Acta Bioeng. Biomech. **18**(2), 121–129 (2016)
2. Kim, C.J., Son, S.M.: Comparison of spatiotemporal gait parameters between children with normal development and children with diplegic cerebral palsy. J. Phys. Ther. Sci. **26**(9), 1317–1319 (2014)
3. Johnson, D.C., Damiano, D.L., Abel, M.F.: The evolution of gait in childhood and adolescent cerebral palsy. J. Pediatr. Orthop. **17**(3), 392–396 (1997)
4. Rodda, J., Graham, H.K.: Classification of gait patterns in spastic hemiplegia and spastic diplegia: a basis for a management algorithm. Eur. J. Neurol. **8**(Suppl. 5), 98–108 (2001)
5. Ross, S.A., Engsberg, J.R.: Relationships between spasticity, strength, gait, and the GMFM-66 in persons with spastic diplegia cerebral palsy. Arch. Phys. Med. Rehabil. **88**(9), 1114–1120 (2007)
6. Kusumoto, Y., Takaki, K., Matsuda, T., Nitta, O.: Relation of selective voluntary motor control of the lower extremity and extensor strength of the knee joint in children with spastic diplegia. J. Phys. Ther. Sci. **28**(6), 1868–1871 (2016)
7. Damiano, D.L., Kelly, L.E., Vaughn, C.L.: Effects of quadriceps femoris muscle strengthening on crouch gait in children with spastic diplegia. Phys. Ther. **75**(8), 658–667; discussion 68–71 (1995)
8. Engsberg, J.R., Ross, S.A., Collins, D.R.: Increasing ankle strength to improve gait and function in children with cerebral palsy: a pilot study. Pediatr Phys. Ther. **18**(4), 266–275 (2006)
9. Eek, M.N., Tranberg, R., Zugner, R., Alkema, K., Beckung, E.: Muscle strength training to improve gait function in children with cerebral palsy. Dev. Med. Child Neurol. **50**(10), 759–764 (2008)
10. Thompson, N., Stebbins, J., Seniorou, M., Newham, D.: Muscle strength and walking ability in diplegic cerebral palsy: implications for assessment and management. Gait Posture **33**(3), 321–325 (2011)
11. Simão, C.R., Galvão, É.R.V.P., da Silveira Fonseca, D.O., Bezerra, D.A., de Andrade, A.C., Lindquist, A.R.R.: Effects of adding load to the gait of children with cerebral palsy: a three-case report. Fisioterapia e Pesquisa 21(1),67–73 (2014)
12. Blundell, S.W., Shepherd, R.B., Dean, C.M., Adams, R.D., Cahill, B.M.: Functional strength training in cerebral palsy: a pilot study of a group circuit training class for children aged 4-8 years. Clinical rehabilitation **17**(1), 48–57 (2003)
13. Liao, H.F., Liu, Y.C., Liu, W.Y., Lin, Y.T.: Effectiveness of loaded sit-to-stand resistance exercise for children with mild spastic diplegia: a randomized clinical trial. Arch. Phys. Med. Rehabil. **88**(1), 25–31 (2007)
14. Patchay, S., Gahery, Y.: Effect of asymmetrical limb loading on early postural adjustments associated with gait initiation in young healthy adults. Gait & Posture **18**(1), 85–94 (2003)
15. Lam, T., Wirz, M., Lunenburger, L., Dietz, V.: Swing phase resistance enhances flexor muscle activity during treadmill locomotion in incomplete spinal cord injury. Neurorehabil. Neural Repair **22**(5), 438–446 (2008)
16. Houldin, A., Luttin, K., Lam, T.: Locomotor adaptations and aftereffects to resistance during walking in individuals with spinal cord injury. J. Neurophysiol. **106**(1), 247–258 (2011)
17. Yen, S.C., Schmit, B.D., Landry, J.M., Roth, H., Wu, M.: Locomotor adaptation to resistance during treadmill training transfers to overground walking in human SCI. Exp. Brain Res. **216**(3), 473–482 (2012)

18. Bjornson, K., Graubert, C., McLaughlin, J.: Test-retest reliability of the gross motor function measure in children with cerebral palsy. Pediatr Phys. Ther. **12**(4), 200–202 (2000)
19. Jensen, R.K.: Body segment mass, radius and radius of gyration proportions of children. J. Biomech. **19**(5), 359–368 (1986)
20. Abel, M.F., Damiano, D.L.: Strategies for increasing walking speed in diplegic cerebral palsy. J. Pediatr. Orthop. **16**(6), 753–758 (1996)
21. Tedroff, K., Knutson, L.M., Soderberg, G.L.: Synergistic muscle activation during maximum voluntary contractions in children with and without spastic cerebral palsy. Dev. Med. Child Neurol. **48**(10), 789–796 (2006)
22. Lam, T., Pearson, K.G.: The role of proprioceptive feedback in the regulation and adaptation of locomotor activity. Adv. Exp. Med. Biol. **508**, 343–355 (2002)
23. Lam, T., Anderschitz, M., Dietz, V.: Contribution of feedback and feedforward strategies to locomotor adaptations. J. Neurophysiol. **95**(2), 766–773 (2006)

Author Index

Printed in the United States
By Bookmasters